Viviane Cristiane Costa
Ana Cristina Ferreira

Teaching Knowledge and Inclusive Mathematics Education

Viviane Cristiane Costa
Ana Cristina Ferreira

Teaching Knowledge and Inclusive Mathematics Education

Investigating the potential of an extension course aimed at teaching mathematics to the deaf

ScienciaScripts

Imprint

Cover image: www.ingimage.com

This book is a translation from the original published under ISBN 978-613-9-61610-7.

Publisher:
Sciencia Scripts
is a trademark of
Dodo Books Indian Ocean Ltd. and OmniScriptum S.R.L publishing group

120 High Road, East Finchley, London, N2 9ED, United Kingdom
Str. Armeneasca 28/1, office 1, Chisinau MD-2012, Republic of Moldova, Europe
Printed at: see last page
ISBN: 978-620-7-88421-6

Table of contents:

Dedication

I dedicate this work to the memory of my dear friend Sandra Helena Pereira Pinto.

Thanks

To God, for the opportunity to live this moment.

To my family, especially my husband and daughter, who supported me and understood the times I was away.

To Professor Ana Cristina, for her competence and dedication. Thank you for the moments I spent together and for the learning you gave me.

To the members of the panel, Lulu Healy and Milton Rosa, for their careful reading and contributions.

To my fellow Master's students for their friendship, moments of exchange and relaxation.

And to all those who contributed in some way to the development of this research.

SUMMARY

Even with the advances in Brazilian legislation related to Inclusive Education, the daily practice of schools and teacher training courses is still a long way from what is desirable. In the specific case of teaching mathematics to deaf students, despite the obligation to have interpreters in schools and the subject of Libras in degree courses, the scenario is not very different. Training teachers to work in this scenario is an urgent task, both in degree courses and in spaces dedicated to continuing training. The purpose of this qualitative research was to analyse the possible contributions of an extension course to the mobilization of teaching knowledge related to the inclusion of deaf students in Mathematics classes in regular classrooms. The course, developed in partnership with the Ouro Preto Regional Teaching Superintendence, took place in five three-hour meetings and involved 20 participants, including teachers, future teachers and Libras interpreters. The data was collected by means of audio or video recordings of the meetings, a field diary and a memorial constructed by the researcher, as well as records produced by the participants throughout the course and questionnaires answered by them. The results, in general, show that all the participants got involved - with greater or lesser intensity - in the activities carried out during the course and that they all became more aware of inclusion, broadening their perception of what a deaf student is capable of learning, but also of the possibilities of teaching mathematics to classes involving deaf students. The analysis was structured from two perspectives on the process: that of a teacher who took part in the course and that of the trainer/researcher. There are indications that both, in different ways, mobilized teaching knowledge for inclusion, especially pedagogical content knowledge. The first, despite being experienced, broadened her knowledge by experiencing and constructing mathematical tasks that could be developed with all students, including deaf students. She also reframed her perception of the roles of the interpreter and the math teacher in inclusive classes. The second, when reflecting on the process of learning to research and be a trainer, without losing sight of her professional experience as a math teacher, noticed an intense mobilization of knowledge arising not only from her studies and preparation for the extension course meetings, but also from contact with the group. This experience profoundly broadened her perception of teaching from an inclusive perspective and of the importance of self-training for teachers and trainers. Finally, the study generated an educational product, aimed mainly at teachers, future teachers and teacher trainers, in which proposals are presented for training teachers to teach mathematics from an inclusive perspective.

Keywords: Inclusive math education, teacher training, teaching knowledge, teaching math to deaf students.

A personal journey towards mathematics education that includes the deaf

My[1] interest in inclusion arose when I was teaching mathematics to deaf students[2] who were not oralized[3] and who communicated using the Brazilian Sign Language (Libras), in a class of hearing students[4] in Youth and Adult Education (EJA) at a state school in Barbacena/MG. Due to my lack of experience in this context, the first contact was difficult. The deaf students were very reserved, preferring to interact only with the Libras interpreter or with other deaf students who communicated in the same way. When the Libras interpreter[5] , for whatever reason, didn't show up for class, the deaf students wouldn't open their notebooks and would just stay in the classroom and keep to the timetable.

Based on these observations, I asked myself: is the presence of a Libras interpreter "Inclusion" itself? Should we accept that we are from different realities and will never be able to interact? Because it's Inclusion, does the way of assessing need to be different? In short, can the math teacher contribute to the inclusion of deaf students?

The principal, supervisor and other teachers at the school didn't have any answers to my questions either. So I asked the Libras interpreter for help. Through informal conversations, I clarified a few points about the teacher's attitude in the classroom and how deaf people communicate through this language (greetings, the alphabet, numbers, signs for operations and basic everyday things). The partnership with the Libras interpreter met my initial demands and brought me closer to the deaf students. I began to promote interaction between these deaf students and the other hearing students.

There was so much enthusiasm that one of the deaf students brought a printout of the alphabet in Libras and distributed it to the whole school. Everyone became sensitized and involved in helping to improve the school environment for these deaf students. The help of the Libras interpreter was fundamental in mobilizing me to understand how deaf students learn mathematics.

I began to make small changes to my classes using the resources available (blackboard, chalk, multiplication tables and educational games) and I noticed that this allowed deaf students to participate even in the absence of a Libras interpreter.

My partnership with the Libras interpreter continued to bear fruit until the end of the school year. At the beginning of the following year, he chose to work at another school and I was left without this support for a few months. So I made an effort to find out more about communicating with deaf people and to modify my lessons. The content written on the board was adapted, prioritizing the use of infinitive verbs. Adjectives, nouns and verbs were highlighted during explanations on the blackboard. My way of explaining the content and preparing the assessments adapted to the new reality, as I no longer had a Libras interpreter.

Despite the difficulties, acceptance was immediate: all the students in the class agreed to collaborate. The hearing students didn't mind when the subjects were explained with a lot of gestures and in a slower way, drawing attention to the key words of the content. As it was an EJA class, the changes in strategies also helped the hearing students' learning. The students and I managed to create an interesting environment in which the interaction between the math teacher, deaf students and hearing students flowed in a way that facilitated the teaching and learning of math.

The deaf students were no longer isolated from the hearing ones. They did math activities together and interacted in such a way that many became friends, overcoming the communication barrier. Through these experiences, I found some answers to the questions that arose at the beginning of this story. I realized that, according to the *Guide to Special Education in the Minas Gerais state education network*, the Libras interpreter, despite their importance in communicating with deaf students, are not responsible for teaching, but rather assist in the teaching and learning processes of

---

[1] The introduction is written in the first person singular because it is a personal experience. However, most of the text is written in the first person plural because I believe that its construction also involved the supervisor of this research.

[2] In this study, I preferred to use the word "deaf" in lower case because it refers to a particularity (difference) of the individual, and not to deaf culture.

[3] Deaf people who can't read lips.

[4] Students without hearing loss.

[5] Professional who helps deaf students communicate in the classroom.

mathematics. This means that the teacher cannot leave it up to them to teach the deaf student. We need to realize that the teacher's role remains the same and that the Libras interpreter is a partner.

Little by little, I learned to plan my lessons taking into account the particularities of the students and trying to respect their differences. In this sense, like Souza (2013), I believe that math teachers can contribute to the teaching and learning of mathematics for deaf students if they prepare themselves for it.

Talking to other math teachers, my experience seemed far removed from what they believed was possible. For many, the inclusion of these students in school was not part of their remit.

So, when I started my Master's Degree in Mathematics Education, I sought to deepen my studies on School Inclusion, Mathematics Education and Teaching Knowledge in order to investigate possibilities for training (initial or continuing) for teachers (or future teachers) who teach (or will teach) Mathematics.

In this sense, I posed the following research questions: *How does participation in an extension course focused, among other things*[6] *, on the inclusion of deaf people in mathematics classes, contribute to the mobilization of teaching knowledge on the part of a mathematics teacher? How does this experience contribute to the mobilization of teaching knowledge on the part of the teacher trainer?*

The more general purpose of the research is to understand the potential of training actions of this nature to mobilize the teaching knowledge of teachers and future teachers. We are interested in investigating possible contributions to the mobilization of teaching knowledge from an extension course which, among other proposals, aimed to raise participants' awareness of the issue of school inclusion through experiences and reflections related to the teaching of mathematics to deaf students.

We use[7] the term mobilization not necessarily in the sense of promoting new learning, but of "setting in motion" knowledge that already exists or is under construction. Each participant takes advantage of this "movement" in their own way, either by effectively incorporating it into their existing knowledge, or by being "touched" by the new ideas (which can lead to a search for a deeper understanding of these ideas, from which new learning can emerge), or by just getting involved in the experience for a moment and then leaving it to one side.

The dissertation is structured in six chapters. The first two chapters represent our efforts to draw up a summary of the educational history of deaf people, public policies and the main studies carried out. Our intention was to build a theoretical perspective that would underpin the methodology and data analysis.

In the third chapter, we outline the methodological approach used, describing and justifying the methodological options. We situate the context and participants of the research, as well as the tasks and the dynamics of their development.

In the fourth chapter, we describe the meetings of the extension course, in order to provide a more in-depth view of the process and the data produced. In the fifth and sixth chapters, respectively, we analyze the trajectory of a participant and the trainer of the extension course.

To conclude, we have made some observations about the study, trying to answer our research question and reflect on the process experienced by the trainer/researcher. We end the text with bibliographical references, appendices and annexes.

---

[6] The purpose of the extension course analyzed in this research was to explore inclusive practices related to teaching mathematics to deaf and blind students. However, our section deals with the moments in the course when the focus was on teaching mathematics to deaf students.

[7] From this point on, writing in the first person plural will predominate in order to represent the collective construction of this research.

Chapter 1

Teaching the deaf from an Inclusive Mathematics Education perspective

> *The world turns and, in these turns, it changes, and in these mutations, sometimes drastic, sometimes not so much, we also get involved and live with the new, even if we don't realize it. There are, however, those who are more sensitive, those who are ready, 'plugged in' to these twists and turns and who give the first cries of alarm, when they foresee the new, the need for the new, the emergence of the new, the urgency of adopting it, so as not to succumb to death, the degradation of time, the decrepitude of life* (MANTOAN, 2003, p.10).

From antiquity to the present day, deaf people have been viewed by society in different ways. In some civilizations, they were a burden on the family and a nuisance to society (LACERDA, 1998), and could be considered divine envoys or evil creatures (SALES, 2013). When they weren't banished or offered up as sacrifices, they became an amusement for the nobility in the role of court jester (MARÇAL SALES, 2009). Speculation about the education of the deaf began to change in the mid-sixteenth century under the influence of the Catholic Church. However, although they came to be seen as beings who had feelings and needed charity and piety, they had to live in special institutions, monasteries or convents (SALES, 2009). Gradually, small movements, initially concentrated in Europe and the United States, began to admit that deaf people could learn if the sounds of speech or ideas of thought were represented in writing (SILVA, 2006).

According to Sacks (2007 *apud* OLIVEIRA 2005), Charles Michel de L'Épee, a French abbot, by associating signs with pictures and written words, realized that he could teach a deaf person to read. In 1755, L'Épee set up a school for the deaf in his own home. It was the first to receive public aid. He invested in training teachers for the deaf and published a book disseminating his techniques (1776). By the time of her death in 1789, L'Epée had already set up 21 schools for the deaf in France and Europe, which revealed deaf intellectuals (OLIVEIRA, 2005). His intense work generated a movement for the education and emancipation of the deaf that spread to various parts of the world (NEVES, 2011).

In Brazil, in 1857, at the invitation of King Pedro II, Hernest Huet (a French deaf teacher and supporter of L'Epée) founded the first school for deaf children in the country: the Imperial Institute of Deaf Mutes, currently known as the National Institute of Deaf Education (INES) (OLIVEIRA, 2005). According to Bueno (1993) and Mendes (2001), with the creation and implementation of the Institute for Deaf Mutes and other institutions focused on other disabilities, a new type of school education began: Special Education. However, while Brazil's Special Education institutions only served to help those in need, based on exclusion and segregation, the Parisian ones remained as workshops (SANTOS; TELES, 2012).

It was only in 1973 that the concept of Special Education began to change. With the creation of the National Center for Special Education (CENESP), the first training courses for teachers in the field of Special Education were set up. In 1985, the Brazilian government created a committee to plan, supervise and draw up policies for the "disabled" (the term used at the time); the following year, a National Coordination for the Education of People with Disabilities; and, in 1990, it implemented a Special Education policy through the National Secretariat for Basic Education (SANTOS; TELES, 2012).

1.1. Public policies related to deaf education

The Brazilian education system, through Special Education and Education policies, offered two types of service: the Regular School[8] and the Special School. Students attended one or the other (ALONSO, 2013). With the development of Special Education, a new vision of education for people with special needs emerged. In 1994, the Spanish government, in partnership with the United Nations Educational, Scientific and Cultural Organization (UNESCO), organized the World Conference on

[8] Schools for students without physical or mental disabilities.

Special Educational Needs, which eventually resulted in the Salamanca Declaration. This document established rules and standards on equalizing opportunities for people with disabilities, demanding that states ensure the education of these people as an integral part of the educational system (UNESCO, 1994). In Brazil, two years after the Salamanca Declaration, the National Education Guidelines and Bases Law (LDBEN), in its Chapter V, defined the concept of Special Education as a teaching modality that is part of the Regular School with specialized care according to the disability.

Since then, Brazil's inclusive legislation[9] , which protects the rights of all people with some kind of disability, has been expanded. According to Passos, Passos and Arruda (2013), among the main Brazilian inclusive laws are:

- the National Education Plan (2001) - which deals with the objectives and targets for the education of disabled people;
- the National Guidelines for Special Education in Basic Education (2001) - which provide guidance on the procedures for standardizing the services provided for in Chapter V of the LDBEN;
- the Resolution of the National Council of Education n^0 1 (2002) - which incorporates the National Curriculum Guidelines for the Training of Basic Education Teachers at higher education level, full degree courses;
- the National Policy on Special Education from the Perspective of Inclusive Education (2007) - which encourages access, participation and learning for disabled students at all school levels; and
- Decree No. 6.571 (2008) - which regulates and specifies specialized educational assistance in regular schools. The School Censuses began to observe that each year there was an increase in the number of students enrolled who needed specialized care.

In 2015, Law No. 13.146 (BRASIL, 2015) was passed, establishing the Brazilian Law for the Inclusion of People with Disabilities (Statute of People with Disabilities). The School Censuses began to show a growing increase in the number of students enrolled who needed specialized assistance, which meant that the policies mentioned only guaranteed the right of access, participation and learning for students with special needs in regular schools (PASSOS; PASSOS; ARRUDA, 2013). With the entry of these students, needs emerged such as the creation of classrooms to cater for them and qualified teachers (BARRETTA; CANAN, 2012).

Gradually, inclusive laws aimed at the needs of deaf people began to emerge. According to Borges (2013), they began to protect rights that only people who are deaf or in this context can enjoy:

- Law No. 10.098 (2000) - which influences the implementation of sign language in deaf education as a first language to eliminate communication barriers;
- the National Program to Support the Education of the Deaf (2001) - which disseminates sign language, trains professionals to educate deaf people who know sign language and discusses theoretical aspects of deafness.
- Federal Law^0 10.436 (2002) - which recognizes the Brazilian Sign Language (Libras) as a legal means of communication and expression for deaf people; and
- Decree No. 5.626 (2005) - which measures the inclusion of the teaching of Libras not only in teacher training courses at secondary and higher education levels, but also for future speech therapists, doctors and other professionals who come to participate in this context.

In Brasilia, a report on the Bilingual Education Policy - Brazilian Sign Language and Portuguese Language - by the Ministry of Education's Working Group in the Secretariat for Literacy and Diversity, designated by MEC/SECADI Ordinances No. 1,060/2013 and No. 91/2013 (BRASIL, 2013), highlighted the guarantees of constitutional and infra-constitutional rights achieved for deaf education and the importance of a Bilingual School[10] :

> the deaf must be linked to a linguistic-cultural education and not to a special education marked by the definition of deafness as a sensory lack, as an anomaly to

---

[9] Legislation providing for the educational rights of people with disabilities.

[10] Schools where the language of instruction is Libras and Portuguese is taught as a second language after the acquisition of the first language (BRASIL, 2013).

> be rehabilitated or corrected by surgical attempts. The Bilingual Schools for the Deaf are specific and differentiated and their criteria for selecting and enrolling students is not disability, but the linguistic-cultural specificity recognized and valued by the Convention on the Rights of Persons with Disabilities, with a view to promoting the linguistic identity of the deaf community, as well as fostering their social development (BRASIL, 2013, p.6).

This document goes far beyond conceptualizing a Bilingual School, recognizing Libras as the language of the Brazilian Deaf Community, establishing that videotaped tests in sign language are an appropriate form of evaluation in schools and recommending the adaptation of institutional exams such as Prova Brasil and the National High School Exam (ENEM).

Our intention in citing the Bilingual Education Policy is to expose the various legislations related to the educational universe of the deaf and to recognize that, although our research does not explore Deaf Culture in depth, we believe it is appropriate to explain the various legislations that aim to expand or improve in some way the constitutional and infra-constitutional rights of deaf people. However, in this research, we will analyze the context of the regular school that receives deaf students (blind or with other special needs) and adopt a vision of inclusion that is more in line with the ideas of Mantoan (2003) explained in the next topic.

This shows that the repertoire of inclusive legislation aimed at the needs of deaf people is promising. Unfortunately, however, we are still a long way from true educational inclusion. In the specific case of Mathematics, according to Borges (2013), learning and interaction between deaf and hearing people in Mathematics classes are still not happening properly. The initial and ongoing training of math teachers and Libras interpreters does not cover this inclusion. Both the role of the Libras interpreter is still under construction and the school curriculum has not been adapted and does not consider adequate possibilities for the learning of these individuals (BORGES, 2013).

All of the above confirms the need to provide teachers, future teachers and other professionals working in educational institutions with opportunities to raise awareness of the issue and prepare themselves to act (UNESCO, 1994).

1.2. School Inclusion

In this topic, we will present a brief discussion of school inclusion. Initially, we will deal with Inclusion in general terms, exploring its principles. Next, we will particularize the discussion, exposing the School Inclusion framework chosen to support our studies.

Like Sassaki (2002), we believe that: "Building a truly inclusive society also involves taking care with language. Language expresses, voluntarily or involuntarily, respect for or discrimination against people with disabilities" (p.1).

From a legal point of view, the term disabled or disabled person, used in Brazil's Federal Constitution (1988), also refers to people whose abilities are temporarily limited (such as someone who suffers a car accident and is unable to walk for some time) (NETO, 2012). However, Sassaki (2002) questions the use of these terms: "a person only carries something that they may not carry, deliberately or casually. For example, a person can carry an umbrella if they need to and leave it somewhere because they forget or decide to. You can't do this with a disability, of course" (p.15).

Like Neto (2012), we understand that "a person with a disability is a person with special needs" (p.15). We therefore prefer to refer to people with disabilities as people with special needs:

> When we talk about special needs, the intention is, in some cases, to generate protection and equality for a larger group of people. In specific cases, not only is the concept of disability restricted, but the type and degree of disability is also delimited (NETO, 2012, p.4).

In the case of deaf people, not all of them consider themselves disabled. Some have their own way of understanding and modifying the world in order to make it more accessible and liveable through their visual perceptions, which make up their deaf identity. This includes the language, ideas, beliefs, customs and habits of deaf people (STROBEL, 2009).

After these clarifications, let's begin our discussion. According to Sassaki (2002), the development of education for people with special needs has gone through phases - exclusion, institutional segregation, integration and inclusion:

> Exclusion phase - In this phase, no educational attention was paid to people with disabilities. They were considered unworthy of school education.
> Institutional segregation phase - As disabled children and young people were unable to access ordinary schools, their families joined forces to set up special schools. Hospitals and homes were also used as places for special education.
> Integration phase - the most able children and young people were sent to ordinary schools, special classes and resource rooms.
> Inclusion phase - All people are included in ordinary classrooms. Physical environments and educational procedures are adapted to accommodate the diversity of students. Schools take into account the needs of all students (p.3).

School inclusion aims to build a school that is open to all people, based on the following principles: celebration of individual differences, the right to belong, valuing human diversity, the contribution of each student to cooperative learning, humanitarian solidarity, the equal importance of minorities in relation to the majority and citizenship with quality of life (SASSAKI, 2002).

We can see that Sassaki (2002) organizes the development of education into phases, giving us the impression that as one phase passes, another begins. However, after reviewing the literature in this work, we realized that we cannot identify the chronological order of all the phases, because each school develops according to its own reality, which can be in any of the phases mentioned by Sassaki (2002).

In these guidelines, as well as in inclusive legislation, the principles that guide school inclusion do not in themselves guarantee that inclusive schools will be effective. In our country, the notion of school inclusion is still treated in a reductionist or distorted way, and this ends up diverting us from the challenges of effectively changing our purposes and practices (MANTOAN, 2003). This view (or distortion) began with the Salamanca Declaration (1994), when the term Integration came to be understood as Inclusion, in the sense that the school would welcome all students, regardless of their particularities. This strengthened the trend towards Integration as Inclusion in education systems (NEVES, 2011). The terms Integration and Inclusion are used to express situations of inclusion; however, they are based on divergent theoretical and methodological positions. We therefore need to differentiate between them in the educational context (MANTOAN, 2003).

Integration refers to the inclusion of people with special needs in mainstream or special schools (in special classes or not), leisure groups or homes for people with special needs. Its aim is to integrate a previously excluded pupil or group of pupils. Through school integration, students with special needs have access to schools through various possibilities. They can move through the school system from mainstream to special education. From the regular classroom to special education in all its forms: special schools, special classes in ordinary schools, itinerant teaching, resource rooms, hospital classes, home teaching and others. In school integration, not all students with special needs are able to attend mainstream classes, as there is a prior selection of those who are suitable for integration. From the perspective of integration, some students would receive differentiated treatment - probably in special schools - with adapted curricula, special assessments and reduced educational objectives to compensate for learning difficulties. This would be a juxtaposition of special education with regular education, causing a swelling of this modality, due to the displacement of professionals, resources, methods and techniques from special education to regular schools (MANTOAN, 2003).

In our view, integration refers more to the socialization of people with special needs. Students with special needs are inserted into the school context, they go to school, they meet other people, but they are not assessed on what they are capable of doing, nor do they establish a commitment to the school, which makes it more difficult to verify whether this socialization adds anything to the student's life with any progress in school knowledge. The school accepts them, but is not committed to their particularities, let alone their progress in learning (MANTOAN, 2003).

For Mantoan (2003), Inclusion questions policies, the organization of special and regular education and the concept of Integration, and envisages school integration in a radical, complete and systematic way. Radicalism, in this case, demands a change in the educational paradigm and seeks to eliminate the subdivision of school systems and modalities: special or mainstream. All students, without exception, must attend mainstream classrooms. Its aim is to leave no one out of mainstream education from the start of their school life and to bring about an improvement in the quality of teaching. The education system in Inclusive Schools involves an organization based on the needs of all students. And such organization implies a change in educational perspective, as it must reach not only students with special needs, but also all those who have difficulties in learning. These schools cater for differences without discriminating, without working separately with some students, without establishing specific rules for planning, learning and assessment (curricula, activities, assessment of learning for students with disabilities and special educational needs) (MANTOAN, 2013).

> Inclusion is an innovation whose meaning has been much distorted and polemicized by the most different educational segments. It is a worldwide movement in which people with disabilities and their families fight for their rights and place in society. It is linked to all people who do not have the same opportunities within society (DELLANI; MORAES, 2012, p.3).

After studying School Inclusion, especially Integration and Inclusion, combined with studies on public policies related to school inclusion, we realized that the distinction between Integration and Inclusion is a step towards achieving a process of transformation in schools, so that all students can be welcomed at the different levels of education with their particularities. Thus, in this study, we understand, like Mantoan (2005), that inclusion:

> It's our ability to understand and recognize others and thus have the privilege of living and sharing with people who are different from us. Inclusive education welcomes everyone, without exception. [...] I often say that to be together is to be crowded together at the movies, on the bus and even in the classroom with people you don't know. Inclusion, on the other hand, is being with, interacting with the other (p.24).

According to Mantoan (2003), in order for school inclusion to be effective, it would be advisable for schools to form study groups in order to discuss and understand the educational dilemmas related to this context, in the light of scientific and interdisciplinary knowledge, if possible. It would be interesting if the groups were organized spontaneously by the teachers themselves during their school hours. The meetings would be based on the common needs and interests of some teachers, in order to clarify situations and improve the way they work in the classroom.

The focus of this in-service training would be the development of competence in solving pedagogical problems. Based on discussions about their own reality, teachers could analyze and evaluate how teaching is being carried out and how students are building their knowledge, and then develop new practices.

We also argue that:

> The central axis of the inclusive proposal is to provide better learning conditions for all through a radical transformation of the pedagogical culture. This requires that interpersonal relationships and pedagogical practice be discussed, thus preventing them from being camouflaged or projected onto the student, who is most often blamed for school failure due to his or her deficiencies (GARCIA, 2006, p.27).

According to Sassaki (2002), the world is moving towards building an increasingly inclusive society. And the development of this process can be seen with increasing frequency in the media, in our neighborhoods, in community resources and in programs and services. For this author, although the results are still small, they are growing and encouraging. There is great determination on the part of some principals, teachers and parents, as well as some state and municipal education secretaries, to learn about and apply the philosophy and methodology of school inclusion:

> The results are still small, but growing and encouraging. With each passing day, I learn of yet another group of people in this immense country wishing to learn about and apply the philosophy and methodology of school inclusion, based on the assumption that all young people and children, with or without disabilities, have the right to study together in order to grow up as happy citizens capable of contributing to improving society's quality of life. The growing inclusive movement is quite visible, fueled by the adhesion of sectors of society (schools, associations, companies, government bodies, specialized institutions, the media, etc.) to the principles of social inclusion. [...] At the same time, a wide range of measures are

> being implemented to adapt school systems to the needs of students, including the elimination of architectural, attitudinal and communication barriers, changes in teaching and learning strategies, the application of the theory of multiple intelligences, adaptations to teaching resources and the introduction of technologies (assistive, computer, locomotion, etc.) (SASSAKI, 2002, p.6).

Like Aranha (2004), we understand that a society can be considered successful when it favors the areas of human coexistence and respect for diversity, guaranteeing all its citizens, at all stages of their existence, the conditions for a dignified life of physical, psychological, social and economic quality. In this context, education plays a fundamental role, as it must favor all people (with special needs or not), promoting access to knowledge and the development of skills. The different curricular contents must be organized in such a way as to make learning effective. The school needs to be organized in such a way as to ensure that every pedagogical action results in a contribution to each student's learning process.

Thus, based on the premises of School Inclusion:

> An inclusive school is one that guarantees quality education for each of its students, recognizing and respecting diversity and responding to each one according to their potential and needs. Thus, a school can only be considered inclusive when it is organized to favor every student, regardless of ethnicity, gender, age, disability, social condition or any other situation. Meaningful teaching is that which guarantees access to the systematized body of knowledge as resources to be mobilized (ARANHA, 2004, p.7).

It is true that in order to become inclusive, a school needs the conscious and responsible participation of all those who make up this educational setting (managers, teachers, students, family members and members of the school community) where each student lives, because:

> Inclusion requires more than laws. It requires adequate attention. Providing materials, resource rooms or specialized teams that visit schools are necessary, but not sufficient. The problems arise on a daily basis, in class, and transcend this narrow scope, reaching the responsibility of the teaching team. The promised institutional support is not enough without the effective participation of the student and, above all, the teacher (FERNANDES; HEALY, 2007, p.73).

In this context, both the construction and implementation of inclusive schools require changes in habits and transformations in ideas, attitudes and the practice of social relations in the political, administrative and didactic-pedagogical spheres (ARANHA, 2004).

From this perspective, we understand that

> Inclusion is a complex process that has different dimensions: ideological, socio-cultural, political and economic. The relational determinants include interactions, feelings, meanings, needs and practical actions, while the material and economic determinants make school restructuring possible. In this line of thinking, inclusive education must start from everyday life: the collective, the school and the ordinary classroom, where all students with educational needs, whether special or not, need to learn, have access to knowledge, culture and progress in personal and social terms (GARCIA, 2006, p.14).

Garcia's (2006) considerations show that school inclusion is a process that develops from day-to-day life, in a collective way, and aims to promote schooling for all students, regardless of their educational needs.

1.3. Teaching mathematics to deaf students: a literature review

We carried out a survey using a parameterized query with the terms "Mathematics" and "Inclusion", in the Directory of Research Groups in Brazil[11] and CNPq, from 21 to 26 January 2015. Seven research groups were found.

In the contact field, we found two groups that had websites, where we searched for theses and dissertations related to teaching mathematics to the deaf that highlighted mathematics teachers. For

---

[11]http://lattes.cnpq.br/web/dgp. The database of theses and dissertations of the Coordination for the Improvement of Higher Education Personnel (CAPES), the most used resource, did not make available the works carried out in 2013 and 2014 despite insistence with *e-mail* messages.

the other groups, we searched the *websites of* their respective institutions, looking for the related *Stricto Sensu* Postgraduate Program. When we found it, we searched its database of theses and dissertations for productions using the words "Mathematics" and "Deaf".

We found 33 studies on the subject of deaf education, carried out in *Stricto Sensu* postgraduate programs. Of these, we selected 20[12] , using mathematics, the level of education of the mathematics teacher and the training of the mathematics teacher as criteria.

After gathering all the research abstracts, we organized the data in a table with author, title, year, advisor, institution, level, purpose of the research, theoretical references used, methodology adopted and results. When the abstracts didn't provide the necessary data, we turned to the full text.

Of the 20 studies, three are doctoral theses and 17 are master's dissertations. The topic still seems to be a recent one in the country, as the first research study was defended in 2005. However, it looks promising, as the number of studies has increased significantly since 2012.

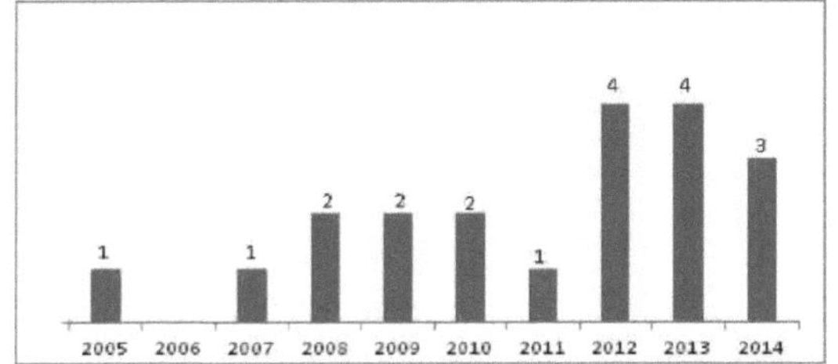

Graph 1 - Number of products per year Source: research collection.

The 20 papers were produced in *Stricto Sensu* Postgraduate Programs in Education, Mathematics Education or Science and Mathematics Education, seven from private institutions and 13 from public institutions.

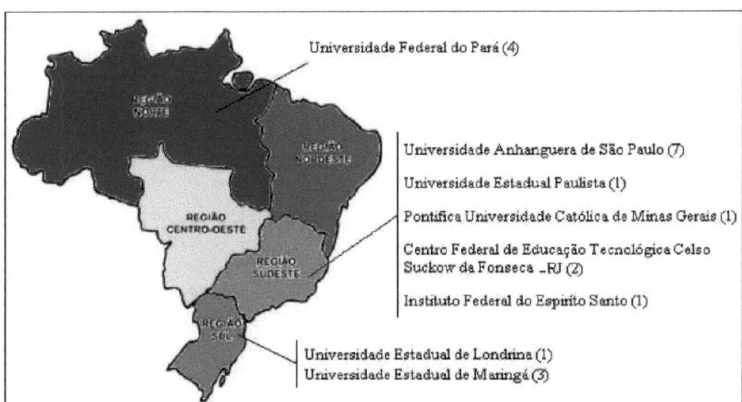

Figure 1 - Location of institutions and number of papers defended at each
Source: research collection.

We noticed a concentration of productions in the Southeast and a small representation in the North and South. The largest number of productions originated at the Anhanguera University of São Paulo[13] , with only two supervisors: Solange Hassan Ahmad Ali Fernandes, with three supervisors, and Siobhan Victoria Healy, with three supervisors and one co-advisor.

The following is just a summary of the review. To find out more about the studies that make up this review, see Appendix 6 (p.144).

[12] The research that was discarded dealt with deaf education, but was directed at the teaching of other content or related to the teaching of mathematics in early childhood education.

[13] Formerly known as the Bandeirante University of São Paulo (UNIBAN).

In the 20 studies related to Mathematics Education for the deaf, there was a predominance of studies on the appropriation of Mathematics topics, the adaptation or insertion of methodologies in Mathematics classes and the training needs of the Mathematics teacher and the Libras interpreter.

Seven of these studies investigate the process of constructing and assimilating mathematical knowledge based on the characteristics of the learners (SILVA, 2008; SALES, 2008; SOUZA, 2010; CONCEIÇÂO, 2012; ASSIS, 2013; JESUS, 2014; FRIZZARINI, 2014). Eleven of them focus directly on the use of pedagogical strategies to promote the teaching and learning of mathematics. Of these works, three use manipulative materials (OLIVEIRA, 2005; SILVA, 2012; JESUS, 2014), five promote inclusion through information technology (SALES, 2009; SOUZA, 2010; CONCEIÇÂO, 2012; SANTOS, 2012; BEZERRA; 2012) and three present adaptations of existing methodologies (SALES, 2013; ASSIS, 2013; SILVA, 2014).

Six studies investigate the training needs of professionals (math teachers and Libras interpreters) who teach math to deaf students (OLIVEIRA, 2005; GIL, 2008; NASCIMENTO, 2009; PAIXÂO, 2010; NEVES, 2011; BORGES, 2013).

In relation to the mathematical content explored in the works, six refer to Geometry (OLIVEIRA 2005; PAIXÂO, 2010; GIL, 2007; SANTOS, 2012; SALES, 2013; JESUS, 2014), three explore concepts and applications of fractions (SOUZA, 2010; ASSIS, 2013; SILVA, 2014), three deal with operations with natural numbers (SALES, 2008; NEVES, 2011; BEZERRA, 2012), three investigate Algebra content (SALES, 2009; CONCEIÇÂO, 2012; FRIZZARINI, 2014), one deals with numerical writing (SILVA, 2008) and one studies the concept of matrices and the addition operation in this context (SILVA, 2012).

Only two studies show interaction between deaf and hearing students during math activities (SANTOS, 2012; BORGES, 2013). In the others, the groups of participants are made up only of deaf people.

Studies show the importance of clarifying educational history, culture and legislation to make math teachers aware of their duties in math education for the deaf (OLIVEIRA, 2005; NEVES, 2011; BORGES, 2013).

Another highlight is the use of Libras as a basis for promoting the teaching and learning of mathematics (NEVES, 2011; ASSIS, 2013; FRIZZARINI, 2014).

The literature studied shows that the interaction between deaf and hearing students is very limited. And the interaction between teachers and deaf students is also far from desirable. As for teacher training (initial and continuing), the studies indicate that it needs to be revised to take into account, among other things:

- the educational history, culture and legislation relating to deaf students; and
- Libras as a basis for promoting the teaching and learning of mathematics.

However, the survey highlights the application of methodologies that seek to contribute to the construction and appropriation of mathematical knowledge by the deaf. Thus, in order to provide teaching and learning situations in mathematics for the deaf, the literature has highlighted the importance of better understanding the educational history, culture and educational legislation in this context. We therefore agree with Souza (2013): there is a need for greater investment in the training of mathematics teachers who teach the deaf so that they can collaborate in the construction of curricula aimed at inclusive schools, as determined by current educational legislation.

Chapter 2
Teaching Knowledge for Inclusion

> *The teaching profession has maintained a very strong individualistic matrix. Unlike other professions that have evolved in a more collective direction - doctors, engineers, architects, lawyers, etc. - teachers have not yet found all the ways to cooperate and collaborate. This is certainly due to the organization of schools and, above all, the organization of teachers' schedules. But I am convinced that this step towards a more cooperative and collaborative profession is decisive for our future* (NÓVOA, 2012, p.110).

In this chapter, we briefly present some public policies related to teacher training in Brazil with regard to inclusion, as well as some authors' views on such training. In addition, we carry out a brief study of teaching knowledge in an attempt to build an understanding of teaching knowledge for inclusion, more specifically the inclusion of deaf students.

The National Education Council's Resolution No.^{0} 1 (2002) incorporated the National Curriculum Guidelines for the Training of Basic Education Teachers, at higher education level, with a full degree (PASSOS; PASSOS; ARRUDA, 2013), with a view to inclusive schools. This Resolution highlights the importance of training professionals who act critically and use different technological resources in order to express themselves in various languages. It is also emphasized that teacher training should allow for the acquisition of knowledge about the peculiarities of students according to their disabilities (CINTRA, 2014).

In July 2015, the National Education Council, in the National Curriculum Guidelines for initial training at higher education level (degree courses, pedagogical training courses for graduates and second degree courses), defined in art. 5, items V and VII:

> V - the development of teacher training processes in line with educational and social changes, accompanying the gnosiological[14] and epistemological transformations of knowledge;
>
> (...);
>
> VIII - the consolidation of inclusive education through respect for differences, recognizing and valuing ethnic-racial, gender, sexual, religious and age group diversity, among others; (BRASIL, 2015).

And for continuing education, it proposes in Article 8° , in sections II, VII and VIII:

> II - to understand their role in the education of primary school students based on a broad and contextualized conception of teaching and their learning and development processes, including those who did not have the opportunity to attend school at the proper age;
>
> (A ....
>
> VII - to identify socio-cultural and educational issues and problems, with an investigative, integrative and propositional approach to complex realities, in order to contribute to overcoming social, ethnic-racial, economic, cultural, religious, political, gender, sexual and other exclusions;
>
> VIII - demonstrate awareness of diversity, respecting environmental-ecological, ethnic-racial, gender, generational, social class, religious, special needs and sexual diversity differences, among others; (BRASIL, 2015).

The aforementioned sections prioritize training that integrates educational and social changes in order to consolidate Inclusive Education, highlighting the importance of teachers in the teaching and learning processes of all their students. They also encourage an investigative, interactive and motivating attitude towards the various situations in the school context in order to avoid any kind of exclusion.

---

[14] Theory of human knowledge: theory that turns to a reflective analysis of the origin, nature and essence of cognitive action, the act of knowing, human knowledge (meaning available at: <www.michaelis.uol.com.br>. Accessed on: June 3, 2016).

As for people with special needs, in the same period, on July 6, 2015, Law No. 13.146 - referring to the Statute for People with Disabilities - was sanctioned in Brazil. Title II of this statute deals with fundamental rights, and Chapter IV deals with educational rights. This chapter reaffirms all the rights defended by the laws described above, emphasizing the different needs according to each type of disability. For the education of deaf people, this law proposes bilingual education and the compulsory use of Libras. In relation to Teacher Training, we find, in art. 28, item X: "adoption of inclusive pedagogical practices by initial and continuing teacher training programs and provision of continuing training for specialized care" (BRASIL, 2015).

Mantoan (2003) points out that initial and continuing training aimed at school inclusion requires a *re-signification of* the role of the teacher, the school, education and pedagogical practices in order to overcome the exclusionary context of our education at all levels. The author also states that preparing teachers to teach in inclusive schools involves a different *design* from existing professionalization proposals.

According to the author:

> We have learned to teach according to the hegemony and primacy of academic content and, naturally, we find it very difficult to let go of this apprenticeship, which holds us back in the processes of re-signifying our role, whatever the level of teaching at which we work.
>
> But are we really sure that our role is the same: that of transmitters of closed and fragmented knowledge, in times and school subjects that imprison us in curricula? Are we so reduced to mere instructors, who lead and guide our students' ability to know, transforming them into passive beings who are comfortable learning what we define as the truth? Have we consulted ourselves about our greatest educational commitment, whether within ourselves, in our schools or in our corporate organizations?
>
> These fundamental issues need to be more exposed and debated, because it is essential that we are clear about our educational dream, or rather, what we want to achieve when we dedicate hours, days, years of our lives to teaching (MANTOAN, 2003, p.9).

In this way, redefining the role of the teacher involves a process of reflection, a rethinking of teaching practice and educational objectives, so that the teacher is not just a transmitter of knowledge. It implies:

> By questioning their own practice, making comparisons and analyzing the circumstances and facts that cause disruption and/or are responsible for school success, teachers are gradually defining their 'pedagogical theories'. The intention is for teachers to be able to explain what they previously only knew how to reproduce from what they learned in courses, workshops, lectures, exclusively. The proposal encourages teachers to interact regularly with their colleagues, to study together and to be open to collaborating with their peers in the search for pedagogical paths to inclusion (MANTOAN, 2003, p.44).

It would therefore be interesting if teacher training courses aimed at this context presented other ways of preparing these professionals to transform the school, with a view to unconditional openness to differences and quality teaching through the sharing of ideas and ideas inspired by real educational experiences:

> The constant and systematic sharing of ideas, feelings and ideas between teachers, principals and school coordinators is one of the key points of in-service development. This exercise is carried out on concrete experiences, real problems, day-to-day situations that unbalance the work in classrooms - this is the raw material for the changes sought by training (MANTOAN, 2003, p.44).

With regard to deaf students, the attempt to include them in classes of hearing students with the presence of a Libras translator/interpreter may not be successful without the teacher's preparation. It is essential for the teacher to understand the cognitive pathways and sign language, so as not to leave deaf students without pedagogical guidance on each of the curricular themes as a whole. The educational reality of deaf people involves a conscious change, which discusses the reasons behind the resistance and rejection of some teachers in accepting them into inclusive schools (FALCÁO, 2012).

> The inclusion of students with disabilities also requires the inclusion of the teachers themselves, so that they have a favorable environment in which to reflect on their practice and the feelings that the presence of a particular disability gives rise to. It is necessary to invest in building a listening sword for these professionals, so that they can give vent to their feelings, whether they are of love, anger, pain, anguish, frustration or (in)satisfaction, sharing and re(signifying) their experience (DELLANI; MORAES, 2012, p.10).

Like Pimentel (2012), we understand that:

> Educational inclusion requires teachers who are prepared to work with diversity, understanding differences and valuing the potential of each student so that teaching favors learning for all. The absence of this training leads to the phenomenon of pseudo-inclusion, i.e. the mere appearance of the disabled student in the mainstream school, without them being properly included in the learning process. Being enrolled and attending mainstream classes does not mean being involved in the learning process of that group (p.141).

In the research by Fernandes and Healy (2007), the authors point out that they were unable to find a teacher who claimed to be prepared to teach in classes with a student with special needs. Some teachers even recognize that inclusion is a process that requires constant improvement, but they emphasize that they did not receive training to work with these students in their initial or continuing education and that courses aimed at training teachers from an inclusive perspective must be committed to training them to respect the diversity of these students.

In view of the above, what knowledge is involved in teacher training that aims to meet the objectives of inclusive schools?

In the following topic, in order to answer this question, we first seek to better understand the meaning of knowledge. To do this, we present a summary of our readings on some typologies used internationally: Shulman (1986), Gauthier (1998), Perrenoud (2000) and Tardif (2002); and nationally: Therrien (1995), Saviani (1996), Fiorentini (1998) and Pimenta (1999). We end with our personal construction of teaching knowledge for inclusion.

2.1. Teaching knowledge for inclusion: a state in process

In the literature, we find various conceptions of the term "knowledge". According to Cunha (n.d.), there are two philosophical meanings of knowledge: Plato's and Kant's. For Plato, knowledge means a true opinion that is linked to an explanation and a well-founded thought. For Plato, knowledge means a true opinion that is linked to an explanation and a well-founded thought; while for Kant, the same term indicates an objective, sufficient or subjective truth.

The text by Cunha (n.d.) also states that, according to Bombassaro (1992), the notion of "knowing" indicates being able to, "understanding", mastering a technique, "being able to handle", "being able to understand". And, according to Furió (1994 *apud* CUNHA, s/d), knowledge, also known as know-how, is referred to philosophically and classified into three groups:

- Declarative knowledge (descriptive or factual) - expressed in the form of propositions in events or thoughts about a particular concept in a descriptive way.
- Procedural knowledge - related to skills and aptitudes that are manifested through know-how that is stimulated in a propositional way.
- Explanatory knowledge - responds to the why of facts and concepts, emerges in the mastery of theories that give meaning and depth to the two previous knowledges.

At another point, Gauthier (1998 *apud* CUNHA, s/d) states that the notion of "knowledge" has three different conceptions that refer to a particular place: subjectivity, judgment and argumentation. For this author, "knowledge" arises from subjectivity. It is related to any type of certainty generated by rational thought (judgment) and is established through a discursive activity through which the subject tries to validate a proposition or an action (argumentation) by opposing doubt, error and imagination. It thus differs from other types of certainty, such as faith and preconceived ideas.

Tardif (2002) studies the notion of "knowledge" within trades and professions:

> In reality, in the field of trades and professions, I don't think we can talk about knowledge without relating it to the constraints and context of the work: knowledge is always the knowledge of someone who is working on something in order to

> achieve some goal. Moreover, knowledge is not something that floats around in space: teachers' knowledge is their knowledge and is related to their person and identity, to their life experience and professional history, to their relationships with the students in the classroom and with the other school actors in the school, etc. It is therefore necessary to shield it by relating it to these constituent elements of teaching work (p.11).

For this author, the notion of "knowing" encompasses teachers' knowledge, skills, abilities (or aptitudes) and attitudes. It involves knowing how to do things and knowing how to be.

Teaching knowledge

The debate on teaching knowledge emerged on the international scene through research in the United States, France and Canada in the 1980s and 1990s (NUNES, 2001). According to Finoqueto, Santos and Terrazzan (2005), this debate recognized the existence of a knowledge base for teaching and the teaching profession, valued the knowledge built up and socialized by teachers in their educational practice and signalled ways forward for teacher training and professionalization[15] .

Based on this recognition, a number of studies have been developed, using "teaching knowledge" as their object and providing a growth and diversification of this theme with each piece of research (CARDOSO; PINTO; DORNELLES, 2012). The studies have shown that teachers have a central role to play in education. In addition, some questions have emerged about the knowledge learned or constructed by teachers in their training process (initial and continuing) and about the knowledge that encourages teachers to develop new assessment practices that allow them to identify progress, difficulties and possibilities for reconstructing students' learning, which has led to a discussion about teachers' professional knowledge (CUNHA, S/D).

In this study, based on the notion of knowledge defended by Tardif presented above, we will understand professional knowledge as the knowledge involved in practicing a particular profession. In this sense, the teacher is a professional who has professional knowledge made up not of "specific knowledge", but of various types of knowledge (CUNHA, n.d.). In the literature, we find different names with similar meanings for this knowledge: teacher professionals (TARDIF, 2002), teacher knowledge (CUNHA, n.d.) or teaching knowledge (PIMENTA, 1999).

Shulman (1986, cited by MORGADO, 2001; FINOQUETO; SANTOS; TERRAZAN, 2005) considers "teaching knowledge" to be knowledge that structures teaching:

- *Knowledge of the teaching content (or specific content)* - expressed through two types of knowledge, substantive and synthetic. Noun knowledge deals with the mastery of concepts, ideas and phenomena in a particular area of knowledge. And the syntactic refers to the methods that produce new information in field investigations with the aim of validating new knowledge;
- *General pedagogical knowledge* - knowledge transcending teaching content (teaching and learning theories, classroom management and organization);
- *Knowledge of the curriculum* - knowledge of the educational program, teaching materials, the curriculum as a policy in relation to official knowledge;
- *Pedagogical content knowledge* - knowledge built up by the teacher when teaching, based on the specificity of the content. This is considered a new type of knowledge. In this knowledge, the teacher is the protagonist, as they articulate their content in such a way that it is understandable to all their students;
- *Knowledge of the students and their characteristics*;
- *Knowledge of educational contexts* - knowledge of group work, school administration, funding policies, characteristics of communities and cultures;
- *Knowledge of educational aims* - understandings, goals, purposes, philosophical and historical foundations of education.

A few years later, Gauthier (1998 *apud* NUNES, 2001) proposed that being a teacher consists of exercising a craft made up of knowledge aimed at meeting the challenge of professionalization.

---

[15] The notions of knowing and knowledge seem to be used synonymously by these authors. In addition, the etymology of the words suggests that they are very similar in meaning [the word knowledge derives from the Latin *cognoscere*, which means the desire to know, to know (ZIMERMAN, 2012)]. In this research, too, we have chosen not to differentiate between them.

This author sees teaching as the mobilization of various types of knowledge that form a kind of "reservoir" from which teachers draw in order to respond to the specific demands of their concrete teaching situation. The teaching skills he identifies are:

- *Disciplinary knowledge* - knowledge produced by researchers and scientists in the various scientific disciplines or by teachers in the specific context of teaching their subject;
- *Curricular knowledge* - knowledge acquired through school programs or scripts drawn up by publishers and approved by the government;
- *The knowledge of the educational sciences* - knowledge produced in training or in practice influencing teaching performance;
- *Traditional pedagogical knowledge* - knowledge developed through the act of "teaching". In this knowledge, each teacher expresses their understanding of school, teacher and classroom. Determined before any teacher training course;
- *Experiential knowledge* - knowledge that refers to the teacher's experience. It is made up of strategies, tricks and ways of doing things which, although tried and tested, remain a secret;
- *The knowledge of pedagogical action* - the teacher's experiential knowledge that becomes public through the research carried out in the classroom.

We realize that the knowledge of pedagogical tradition and the knowledge of pedagogical action, despite having the word pedagogical in common, have different meanings. The first relates to the teacher's personality (identity, culture and opinions). The second refers to the teacher's practice in the academic sphere, especially in terms of conducting and disseminating research.

Another researcher who has focused on this subject is Tardif (2002). For him, teaching knowledge involves knowledge, skills and attitudes that are mobilized in teaching, and which come from other knowledge. This plural knowledge involves:

- *Knowledge of professional training* (or of educational sciences and pedagogical ideology) - knowledge transmitted by teacher training institutions. Knowledge intended for the scientific or erudite training of teachers;
- *Pedagogical knowledge* - knowledge mobilized in teaching practice is constituted as doctrines or conceptions derived from rational and normative reflections;
- *Disciplinary knowledge* - social knowledge defined and selected by the university and offered in the form of disciplines, corresponding to the various fields of knowledge and emerging from the cultural tradition and social groups that produce knowledge;
- *Curricular knowledge* - knowledge presented in the discourses, objectives, contents and methods of school programs. The school institution categorizes and presents selected social knowledge as models of culture and training;
- *Experiential knowledge* - knowledge developed by teachers themselves in the course of their practice, based on their daily work and knowledge of their environment (TARDIF, 2002, p.36-39).

In Brazil, Therrien (1995 *apud* NUNES, 2001) is one of the first researchers to highlight the importance of thinking about teaching knowledge and teacher training beyond the academic, involving the teacher's personal and professional development (NUNES, 2001). This author emphasizes that "teaching knowledge" encompasses the knowledge mobilized in the teacher's practice and related to professional experience. This knowledge can be transformed and becomes part of the teacher's identity, constituting a fundamental and central element in pedagogical practices and decisions, and representing original knowledge. According to the author, this knowledge is characterized by being originated and validated in the daily practice of the profession, implying a know-how or know-how that directs the teacher's attitude.

Saviani (1996 *apud* BARROS, 2015) argues that teaching knowledge goes beyond the application of specific attitudes and content and also involves knowledge of the right conditions for learning and school development. This author classifies "teaching knowledge" into five categories:

- *Attitudinal knowledge - Knowledge* contained in the teacher's attitudes, postures and experiences, such as: discipline, punctuality, clarity, fairness, equity and dialog;

- *Critical-contextual knowledge - Knowledge* related to understanding the conditions, characteristics and transformations of the society in which students belong and must be integrated as citizens;
- *Specific knowledge - Knowledge of* the contents of school curricula from the sciences: nature, arts, humanities...;
- *Pedagogical knowledge - Knowledge* produced by the sciences of education presented in educational theories with the aim of articulating pedagogical practice. It helps teachers to situate themselves in the performance of their profession;
- *Didactic-curricular knowledge - Knowledge* relating to the organization and implementation of teaching action. It refers to the mastery of content, instruments and procedures belonging to the pedagogical context.

Fiorentini (1998 *apud* NUNES, 2001) used the relationship between theory and practice as a starting point to identify and characterize "teaching knowledge" and how it could be appropriated or produced by teachers through reflective and investigative pedagogical practice. For this author, it is the relationship that academics and teachers have with knowledge that makes the difference. In this way, the articulation between theory and practice can contribute to teacher or researcher training in a continuous and collective way, as long as pedagogical practice is used as an argument for problematizing, signifying and exploring content in theoretical training.

Pimenta (1999) carries out research based on his practice with undergraduate students and highlights the importance of mobilizing the knowledge of experience to build a teacher's professional identity. He shows that the teaching knowledge inherent in teaching is re-elaborated and constructed by teachers "in confrontation with their practical experiences" (PIMENTA, 1999), lived out daily in school contexts. It also identifies three types of teaching knowledge:

- *Knowledge of experience* - knowledge learned by the teacher as a student, acquired from colleagues or teachers, or in a process of reflection;
- *Knowledge of knowledge* - knowledge that encompasses a review of the school's role in transmitting knowledge and its specialties in a contemporary context;
- *Pedagogical knowledge* - knowledge that, together with knowledge of experience and specific content, builds knowledge based on real pedagogical needs (PIMENTA, 1999 *apudNUNES,* 2001, p.34-35).

In general, we observed that teaching practice occupies an important place in the conceptions of patient knowledge defended by the authors studied. Thus, Therrien (1995) highlights teacher practice as a privileged locus, since teaching knowledge emerges as knowledge of experience. This knowledge helps to shape the teacher's identity and becomes fundamental to their pedagogical practices and decisions. Saviani (1996) argues that teaching knowledge goes beyond a knowledge of attitudes and specific content coordinated by the teacher and also requires knowledge of the appropriate conditions for learning and school development.

Fiorentini (1998) explores the link between theory and practice, which can contribute to teacher training so that they reflect on their practice. Pimenta (1999) considers all the aspects mentioned by the Brazilian authors above to be essential. However, for her, the knowledge of experience must also take into account the knowledge and skills learned by the teacher as a student, since it is not isolated knowledge. It develops in teaching practice together with knowledge of experience and knowledge of knowledge.

These typologies highlight the complexity of the teaching profession. We understand that teaching knowledge does not represent something unique or standardized. It is heterogeneous knowledge influenced by social and cultural contexts, based on knowledge acquired by the teacher in his or her student, professional and personal life (TARDIF, 2002) and is constructed, used and validated according to his or her conceptions of the situations that occur in teaching practice. According to Tardif (2002), teachers eliminate what seems abstract and unconnected to lived reality and keep what may be useful to them at some point (TARDIF, 2002).

In addition, teachers bring with them life experience, beliefs, ideologies and doubts about teaching practice. These phenomena stand out in the exercise of the profession over time. In this way, teachers develop their own teaching knowledge influenced by personal, temporal, professional and academic aspects (TARDIF, 2002). This makes us think that teachers can also develop their teaching knowledge influenced by the principles of School Inclusion, a subject we will deal with in the following topic.

2.2. Teaching knowledge for inclusion

Inclusive schools require teachers to have a professional practice that focuses on: coexistence, recognition, questioning differences and learning as a relational, participatory and meaningful experience for all students (MANTOAN, 2013). Thus:

> If a student is not doing well, whether they are disabled or not, the problem needs to be analyzed in relation to the teaching that is being given to everyone else in the class. This is an important indicator of the quality of pedagogical work, because the fact that the majority of students are doing well does not mean that the teaching is meeting everyone's needs and possibilities (MANTOAN, 2013, p.46).

An inclusive teaching practice involves technical, human, social, cultural and political aspects and is permeated by values, interests, needs, conditions and individual specificities.

> [...] the question of knowledge is closely linked to the question of teaching work in the school environment, its organization, its differentiation, its specialization, the objective and subjective constraints with which teachers have to deal, etc. It is also linked to the whole social context in which the teaching profession is inserted and which determines, in various ways, the knowledge required and acquired in the exercise of the profession (TARDIF, 2002, p.218).

Some studies have tried to answer this question. Cunha, Palma, Basckes, Pedroso and Zat (2011) interviewed the teachers of a school that had a positive experience of Inclusive Education. Based on Tardif (2002), they investigated how teachers' knowledge is constructed in the face of legal and social demands related to the inclusion of students with special needs. According to the information provided by the teachers interviewed, the school's pedagogical project was a decisive factor in the construction of knowledge. The school management, based on public policies that established the concept of Inclusive Schools, proposed in-service training at the school itself. In the view of the authors, this allowed the teachers to build knowledge through a relationship between theoretical contributions and the practices they had experienced.

For the authors, the inclusion of students with special needs, in addition to experiential knowledge, encompasses the perception of specific knowledge and a positive relationship between the components of the school community. This brings us to experiential knowledge, curricular knowledge and pedagogical knowledge according to Tardif (2002).

In the same vein, Machado (n.d.) sought to understand what knowledge related to the practice of teachers for the inclusion of students with special needs is and how it can be constructed, based on the narratives of female teachers. The research was based on Mantoan (2003), Edler Carvalho (2004), Gauthier (1998), Tardif (2006) and Osório Marques (2000). The teachers taking part in the study listed their knowledge as: theoretical (disciplinary and curricular), experiential, from the science of education, from pedagogical tradition and from pedagogical action.

According to the author, the construction or acquisition of the knowledge necessary for inclusive practice in relation to students with special needs can originate in courses, seminars, lectures, books and exchanges of experiences between teachers working in mainstream and/or special schools. These practices can take place in a school environment as a forum for discussion and questioning between school and family, in a university environment, through practicums and internships, in families and even on the internet.

Coutinho (2013) used a narrative interview with the aim of identifying, characterizing and understanding how teachers on the Pedagogy course have built up knowledge to work with students with special needs. The work was based mainly on studies by Gauthier (1998) and Tardif (2002). To do this, the researcher organized and analyzed the data based on Bardin's (1977) content analysis. Coutinho (2013) then began the process of categorization, with categories suggested *a priori* by the theoretical framework and with *a posteriori* categories drawn up after analysing the material. In all, 19 pieces of inclusive knowledge were identified, classified into four analytical categories: central knowledge, structuring knowledge, peripheral knowledge and isolated knowledge. In addition to the inclusive knowledge mentioned, the teachers revealed that the construction of specific and theoretical knowledge related to school inclusion takes place timidly during initial and continuing training.

According to Coutinho (2013), although the experiences lived in the classroom and in the various educational contexts contribute significantly to the construction, mobilization and articulation of teaching knowledge for the inclusion of all students, there is a need for greater investment by public policies in higher education. The researcher points out that it would be appropriate to provide training opportunities for everyone in the academic community (teachers, students, managers and technical-administrative staff). And so offer members of higher education the possibility of building knowledge that promotes pedagogical practices that ensure the right of access and permanence of all students in this form of education.

Another study carried out in this context was by Silva (2014). In it, the author discusses the teaching knowledge that teachers mobilize in the pedagogical practice of students with special needs in a regular public school, supported by the studies of Tardif (2002), Pimenta (2005), Freire (2004), Shulman (2005) and Barth (1993) (to support the teaching knowledge developed), Schön (*apud* NÓVOA, 1995) (to support the study of pedagogical practice for students with special needs) and Bourdieu (1998) (in the concepts of *habitus* and *campus*). Data was collected using questionnaires, interviews and observations of teachers' pedagogical practice in the classroom. Four categories of knowledge were then identified: teachers' conceptions and practices of Inclusive Education; inclusive education policies in relation to the pedagogical practice of people with special needs; training and experience as significant spaces of knowledge that integrate the pedagogical practice of students with special needs; and experience as a possibility and a path towards the inclusion of students with special needs.

At the end of her research, Silva (2014) concludes that, although education policies point to advances in the concepts of inclusion for students with special needs, there are still many conflicts that teachers face in realizing the ideas of education. In their speeches, the teachers highlighted limitations in their training to work with students with special needs and presented knowledge that is part of the pedagogical practice that gives these students the opportunity to learn.

For this author, teaching knowledge related to inclusion comes from various sources and contexts with the aim of establishing coherence between pedagogical know-how and know-how in an attempt to develop students' intellectual capacities.

The studies mentioned above have some common aspects. They all mentioned inclusion policies, investigated teachers who were already working in this context and highlighted the importance of building teaching knowledge related to inclusion in initial and continuing training. Tardif (2002) appears as the main theoretical reference for most of them.

Coutinho (2013) and Silva (2014) used different categorization processes. Coutinho (2013) worked with categories constructed *a priori* by the theoretical framework and with *a posteriori* categories developed after analyzing the data. Silva (2014) identified his categories in the process of analyzing the data collected and produced during the research in dialogue with the theories presented in his work.

Supported by their respective theoretical frameworks, the researchers conducted conversations, interviews, narratives, observations and questionnaires. Only Coutinho (2013) observed the teaching practice of his research participants.

The literature review reinforces what we already thought: it is essential to invest in initial and continuing teacher training from an inclusive perspective if we want to effectively transform education. This does not mean that it is enough, but it is a central aspect in the process of building a new educational culture that truly includes all students.

In relation to teaching mathematics to deaf students specifically, we understand that teachers and future teachers need to learn to communicate (at least minimally) with all students, including the deaf, and build teaching strategies that facilitate mathematical learning. This idea is corroborated by Neves (2011) when he states that "unquestionably the process of mathematical literacy with deaf people must involve a profound mastery of sign language, mathematical knowledge and appropriate methodologies that can make learning meaningful for these students" (p.18).

This means that, in addition to learning some rudiments of Libras, the teacher will need to work together with the interpreter. In this sense, like Borges (2013), we believe that:

> The official discourse on school inclusion invites all those who live in schools, whether as teachers or students, to believe that the presence of students with different characteristics in the same room is something inevitable and good. We agree that there can be contributions for everyone when living with marked differences, such as those who communicate orally and those who communicate through sign language. However, it remains for us to reinforce the school's major role, which is to educate, to provide students, whatever they may be, with sufficient school knowledge to live in the most varied social environments (p.176).

Based on the teaching experience of one of the researchers[16] , as well as on our studies, we present below some considerations about the teacher-deaf student relationship in math classes. We would point out that our experience has always involved welcoming deaf students into regular classes in public schools, sometimes with the support of Libras interpreters.

When a teacher arrives in the classroom, they always greet their students orally. In the case of deaf students, it's interesting to look them in the eye and signal: good morning, or good afternoon, or good evening, or simply hello.

Figure 2 - Greetings in Libras
Source: http://slideplayer.com.br/slide/2876137/

When explaining content or presenting instructions, it is essential that the teacher doesn't turn his or her back on the deaf students, as some can lip-read and understand what is being said. To facilitate communication during lessons, the teacher can ask the interpreter for help, either to "translate" the instructions for the deaf students, or to teach them some signs in Libras, which can be useful in the school context, such as the alphabet, numbers and the signaling of some situations that may occur in everyday life (Got it?). Is it easy? Difficult? Is it over? Right. Wrong. Excuse me. Excuse me. Please). "If there is no minimum relationship between the hearing teacher and the deaf student, we will be omitting the obligation of every educator with any of the students served, which is to aim for learning" (BORGES, 2013, p.60).

If the teacher doesn't know how to sign something in Libras, it can be useful to write it on the board one word at a time, using verbs in the infinitive without articles or prepositions. Drawing or projecting pictures to represent the problem situation can also help. Visualization can be a great ally for interpreting and understanding knowledge.

> Both in mathematical and other contexts, visualization is related to the ability to: create, manipulate and 'read' mental images, guiding and assisting in the formation of logical connections and demonstrations; visualize spatial, quantitative information and visually interpret the information presented to them; review and analyze steps previously taken with objects they could touch and draw and interpret (GOLDENBERG *apud* SALES, 2013, p.70).

---

[16] Viviane Cristiane Costa.

However, some mathematical concepts do not have standardized signs in Libras. So the teacher can negotiate a sign with their deaf student.

> One of the first barriers to this process is the fact that, as sign language is still in the process of being constructed, there are many mathematical symbols that do not have an equivalent in signs. Many signs are provisionally agreed between the educator and the student in the classroom so that the teaching and learning process can take place; hence the strong need for the educator to have a thorough command of these two areas of knowledge, as well as knowledge and appreciation of the specific characteristics of this student. This is a situation that we see happening in everyday deaf classrooms (NEVES, 2011, p.20).

As for the assessments, we think it would be interesting for the teacher to give them to all the students, whenever possible[17] , but taking a few precautions that can make it easier for deaf students to read them, such as: writing the question statements in Portuguese, prioritizing verbs in the infinitive[18] and underlining the words that make sense in Libras[19] . Of course, if the teacher wants to talk about various subjects with a deaf student who communicates in Libras, he or she will have to take a specific course and practice a lot. However, when teaching mathematics to deaf students, these recommendations can help them learn.

> Any pedagogical activity that takes into account the specificities of deaf students - with an emphasis, in this case, on a concern for better communication in math classes - will also have a positive impact on other students. This is because it is not only deaf students who can present learning problems when there is no adequate communication and language, but also any hearing student who shares a common language with the teacher (BORGES, 2013, p.177).

These suggestions are aimed at changing the current scenario because, like Borges (2013), we note that "there are still a number of barriers that stand in the way of inclusion aimed at good quality education for all. Even today, we see policies being implemented without reflection, without a substantial change in school practices" (p.176). In this sense, we believe that teaching knowledge for inclusion is built, expanded and mobilized continuously throughout a teacher's career, making up a professional identity that contributes positively to the school life of all students.

In summary, the construction of teaching knowledge for the inclusion of deaf students involves:

- rethinking/expanding pedagogical knowledge in order to meet the needs of deaf students, trying, as far as possible, to think of strategies that involve the whole class;
- reflect/rethink the conceptions about deafness in order to realize that it is not necessarily associated with learning difficulties, and
- reflect/rethink the perception of deaf students (they are not the object of pity and commiseration, but students with as much potential to learn as any other).

[17] Some deaf students may be unfamiliar with fundamental mathematical concepts. Diverse work is therefore necessary. However, whenever possible, it is interesting for deaf students to be able to keep up with the rest of the class.

[18] In Libras, there are no verb conjugations, only infinitive verbs (walk, pick up, count...).

[19] Prepositions, conjunctions and articles have no meaning in Libras, only nouns, adjectives and verbs.

Chapter 3

The study methodology

This research arose from our reading and experiences as a math teacher in public school classes that received deaf students and from our interest in contributing in some way to the learning of deaf students through teacher training. It also arose from the daily difficulties encountered by deaf students and teachers in math classes and our desire to contribute, albeit modestly, to overcoming this scenario.

3.1. Research question and objectives

Based on our reading and reflections on school inclusion, teaching knowledge and the teaching and learning processes of mathematics for deaf students and the difficulties they encounter, we outlined our object of study: the teaching knowledge of teachers who teach mathematics, more specifically the teaching knowledge related to the inclusion of deaf students mobilized in an extension course. We therefore set out to answer the following research questions:

How did taking part in an extension course focused, among other things[20] *, on the inclusion of deaf people in mathematics classes contribute to the mobilization of teaching knowledge on the part of a mathematics teacher? How does this experience contribute to the mobilization of teaching knowledge on the part of the teacher trainer?*

We are interested in understanding the potential of training actions of this nature to mobilize the teaching knowledge of the participants, especially the mobilization of teaching knowledge related to the inclusion of deaf students in mathematics classes in regular classes. In addition, we sought to understand how this process influences the teaching knowledge of one of the teachers who worked as a trainer on the extension course.

As mentioned earlier, the term "mobilize" is understood here as "set in motion". Aware that it would be very difficult to effectively promote teacher learning from an inclusive perspective in so few meetings, we set out to provide opportunities for reflection, experience and the collective production of teaching proposals with a view to raising awareness among the participants. Our aim was to mobilize knowledge related to teaching mathematics to deaf students, showing that it was possible (in other words, deaf students can learn mathematics, teachers with no previous training can learn to teach mathematics to them and all this can happen in regular classes, in which deaf and hearing students share the same sword) and feasible (without necessarily having great costs or needing sophisticated materials). In other words, we aim to provide a first approximation to Inclusive Mathematics Education, with an emphasis on the inclusion of deaf students.

This qualitative study involved the construction and development of an extension course aimed at teachers who teach mathematics in basic education and future mathematics teachers, with a focus on the collective construction of knowledge and inclusive practices related to the teaching of mathematics. This study involved two research projects, developed by two of the main researcher's students, and covered mathematics teaching for the deaf and mathematics teaching for the blind.

In this way, the *corpus* that makes up this research involves the meetings in which the focus was on teaching mathematics to the deaf. Of course, the notion of inclusion and discussions about it and inclusive math education permeated all the meetings. However, for the purposes of this analysis, we will consider the production most directly related to the tasks and proposals developed in the context of teaching mathematics to the deaf.

3.2. Context

In partnership with the 25ª Superintendencia Regional de Ensino de Ouro Preto (SER-OP), we offered an extension course (Inclusive Mathematics Education and the professional development of teachers who teach Mathematics: building/improving knowledge in an extension course) to teachers who teach Mathematics in Basic Education and future Mathematics teachers, students on the Mathematics degree course at UFOP.

The 25th SRE-OP could select up to 20 participants (including primary and secondary school

[20] The purpose of the extension course analyzed in this research was to explore inclusive practices related to teaching mathematics to deaf and blind students. However, our section deals with moments in the course when the focus was on teaching mathematics to deaf students.

teachers and up to five employees of the 25th SRE-OP). Five other places were open to students on the Mathematics degree course at the Federal University of Ouro Preto (UFOP). However, as the 25th SRE-OP did not fill all the vacancies as we had initially proposed, it was decided to invite students from the Teaching Technician course, who were students of a teacher who wanted to take part in the course.

The course took place in a public school in Ouro Preto, at the request of the 25ª SRE-OP. The school was centrally located, making it easy for participants to access.

3.3. Participants

The course was attended by 20 participants. However, only a few people attended all the meetings described in this paper. The first meeting was attended by 15 people, the second by 18, the third by 15 and the fourth by 13.

We present the participants below, taking as a reference the initial questionnaire answered individually, as well as information from the field diary and conversations that took place during the meetings. In order to preserve the identity of the participants, we have chosen to use pseudonyms.

- Lívia, Rosi and Zeca were employees of the Secretary of Education of the State of Minas Gerais. The first two (Lívia and Rosi) had studied Higher Normal Education and Zeca had studied Teaching. They all worked in the Teacher Training Sector. Lívia had less than two years' teaching experience and Rosi and Zeca had more than five years.
- Bela, Nora, Gina, Célia, Cida, Dara, Leina and Dora were students on the Teaching Technician course with an emphasis on Early Childhood Education. They all belong to the same class and are halfway through the course. The course allows them to work as teachers in Early Childhood Education and the initial years of Primary School. They had different professions: Bela was a cab driver, Gina was a monitor at a nursery school and the others were dedicated to their studies.
- Luna, Marta, José and Juca were students on the Mathematics degree course at a federal university. They were at different points in the course and none of them had any teaching experience. Some were at the beginning of the course and hadn't done any internships, but were taking part in the Institutional Program for Teaching Initiation Scholarships (PIBID)[21] . One of the students was finishing her course and had completed all her internships; she had also taken part in PIBID and other university projects. The last participant had completed his first internships in the year of the study and was developing a scientific initiation project on indiscipline in the mathematics classroom. Three of them had taken an elective course in the Mathematics degree course, in which they had developed teaching projects aimed at including blind and deaf students.
- Julia, Mel and Tina were elementary school math teachers. Julia had a degree in Civil Engineering and Pedagogy. Despite her degree, she had been working as a math teacher for more than five years and, at the time of the fieldwork, she was teaching a technical course in teaching. She took the initiative to invite the students mentioned above. Mel and Tina had already worked in Early Childhood Education as teachers and supervisors, and said they had more than five years' experience.
- Mara and Lara were Libras interpreters. In addition to the training required to work as interpreters, both had other qualifications. Mara had graduated from Normal Superior and had less than two years' teaching experience. Lara had a degree in teaching and pedagogy and had between two and five years' teaching experience. ...

3.4. Meeting dynamics

At the request of the 25ª SRE-OP, the course took place on Saturday mornings. There were five meetings lasting three hours (9am to 12pm). During these meetings, a variety of activities were developed, including workshop moments and moments to study texts, videos and reflections built collectively on the inclusion of blind and deaf students. We took part in four of the course's five meetings, as one of them focused exclusively on teaching mathematics to the visually impaired.

---

[21] The program offers teaching initiation scholarships to students on face-to-face courses who work as interns in public schools and who, when they graduate, commit themselves to teaching in the public sector. The aim is to anticipate the link between future teachers and public school classrooms. Available at: <http//:portal. mec.gov.br/pibid> Accessed on: April 13, 2017.

At the first meeting, we presented the course proposal, and the participants had the opportunity to express themselves about it, suggesting the insertion or alteration of themes and tasks:

Date	Activities
14/11/2015	Packaging handling. Reflections on inclusion. *[On that day, there were activities aimed at teaching deaf and blind people. We will highlight here only those related to teaching mathematics to deaf people].*
28/11/2015	Building a Christmas star. Interview with Camila (deaf student). Learning a little about Libras.
05/12/2015	Discussion of the text: "Talking with your hands" (available at https://novaescola.org.br/conteudo/1611/falar-com-as-maos). Activity involving equivalent fractions.
12/12/2015	Presentation of the math teaching proposals prepared by the participants. Course evaluation.

Table 1: Extension course timetable

Each meeting, starting with the second, was planned with the clear aim of promoting the mobilization of teaching knowledge for inclusion in mathematics classes. We sought to create a space for collective reflection in which each participant could develop their own inclusive practices based on this new knowledge. In this way, we used the participants' reflections and evaluations from each meeting to plan the next meeting.

In terms of attendance, we had an average of 15 participants at each meeting. Despite our efforts to ensure that everyone attended the meetings, Mel and Tina only attended the first one, Júlia and Zeca only missed the first one and others only attended one meeting.

A written and oral evaluation was carried out after each meeting with the aim of checking the process experienced each day and looking for new themes/content of interest to the participants. Based on these, we constructed the tasks for the following meetings.

In our research, we took on the dual task of investigating teachers' knowledge for inclusion in a study group and participating in this process as a member of the group. In this way, we acted as both researcher/trainer and participant.

Throughout the fieldwork, we were concerned both with carrying out the research and contributing to the participants' training, making ourselves available to help them in any way we could. In this sense, respect for the participants, for their professional work and for their contributions to the group permeated all the work. We tried to give priority to the exchange of knowledge and the personal and professional growth of each member of the group, so that the participants felt comfortable talking about their difficulties and desires, their personal experiences and expectations in relation to the meetings.

3.4.1 Data collection

The research considered data collected in November and December 2015 in four meetings that specifically addressed the inclusion of deaf students in math classes. The data was collected using different instruments:

I - Survey

We carried out a survey at the beginning of the research, with the aim of identifying the participants' perceptions of school inclusion, in order to guide the tasks to be worked on in the following meetings (see Appendix 1, p.131).

II - Researcher's field diary

Throughout the process, we recorded personal impressions, behaviors, doubts and ideas of the participants in our field diary. This tool was very important for recovering information from each meeting, especially in relation to some of the participants' speeches. The information collected during the fieldwork was immediately rewritten after each meeting.

III - Audio/video recordings

During the meetings, we prioritized recording the activities and oral reflections. Some moments were recorded on audio and video, and others only on video, as some activities did not have audio. All the material collected relating to our object of study was transcribed. From these, we extracted most of the data considered in the research.

VI - Records produced by participants

At the end of each activity, we asked the participants to write an individual reflection (after each activity) and an evaluation (at the end of each meeting). These procedures were carried out either freely or by means of questionnaires (see Appendices 3 and 4, p.135 and 140). In this way, the participants were able to record their perceptions of the work carried out: positive points, aspects for improvement and suggestions for mathematical content. To make them feel more comfortable expressing their opinions without embarrassment, the participants were informed that they did not need to identify themselves.

The words of the participants and the researcher were obtained from field notes and audio and video recordings. Some images were also used to illustrate the development of activities carried out by the group participants. Each piece of information used has been duly referenced. All records produced by the researcher were considered field notes.

VII - Teacher trainer's memoir

At the same time as this research was being carried out, one of the researchers (the author of this dissertation) was the subject of another Master's research project[22] , which analyzed the potential of the construction of training memorials for the professional development of mathematics teachers. When we decided to analyze the experience of the author of this research, who acted as a trainer in the extension course, after the qualification, taking up an interesting suggestion from the Board, this memorial came to the fore. A large part of the reflections recorded in it referred to the researcher's

[22] Dissertation entitled "Stages of self-writing: formative trajectories of mathematics teachers in memorials", defended by lara Leticia Leite de Oliveira, in March 2017, within the framework of the Master's Degree in Education at UFOP. The research involved "6 Mathematics teachers, students in a Postgraduate Program in Mathematics Education, whose collaboration took the form of a questionnaire and the writing of a training memorial that constituted the *corpus of* analysis" (abstract).

experience during her Master's degree when, as well as being a teacher learning how to research, she also had to learn how to act as a trainer in the extension course. We therefore used this material in our analysis. ....

3.4.2 Data analysis

The analysis and interpretation of data in qualitative research is a moment when the researcher seeks to answer the research questions based on an articulation between the data collected and its theoretical foundation (MINAYO, 2011).

Initially, all the data collected was organized and, from there, we built the next chapter in which the process is reported meeting by meeting. To describe the meetings of the extension course, in addition to the questionnaires, reflections and evaluations for each meeting, we used all the audio, video and photo files that were stored on the computer during the fieldwork. Another resource used by the researcher was the field diary, which was also stored in a computer file. In this diary, the researcher made notes on the planning of the activities, the participants' speeches and the development of the meetings.

After describing the process, we defined the analysis. Due to time constraints, we chose to analyze the process experienced by the only math teacher in the group who took part in almost all the meetings (Julia). To do this, we looked at all the data related to her, trying to uncover possible mobilizations of knowledge.

In addition, as mentioned above, after the qualifying exam, we decided to analyze the process experienced by the researcher/trainer, the author of this research.

We hadn't originally planned to analyze the contributions of the extension course in this dimension. However, they turned out to be an interesting and still relatively unexplored avenue of research. We considered all the data collected, as well as the memoir. Once again, we used the field diary, audio files, videos and photos. To this end, we have tried to report once again on the extension course from the trainer's point of view, seeking to shed light on possible mobilizations of knowledge.

Chapter 4

The extension course

In this chapter, we describe the meetings of the extension course. Our aim is to provide an overview of the process and the data produced. The first two meetings will be presented in more detail and the following ones only briefly. However, the interested reader can find more information in the Appendices of this Dissertation and in the Educational Product produced from it[23] .

The course took place in five meetings lasting three hours, on Saturdays, in a classroom at a state school in Ouro Preto. During these meetings, a variety of activities were developed, including moments of workshop and moments of studying texts, videos and reflections built collectively. We asked[24] for an individual written evaluation from the participants at the end of each meeting. They could record their perceptions of the work they had done (positive points, aspects for improvement, etc.). To make them feel more comfortable expressing their opinions without embarrassment, the participants were informed that they did not need to identify themselves.

The data collected in four[25] of these meetings, which dealt specifically with the inclusion of deaf students in math classes, was considered here.

4.1. Meeting on 14/11/15

The first meeting began with a coffee in which we introduced ourselves and the proposal for the course, mentioning that it was part of two Master's research projects and that, with the participants' permission, we would be collecting information. During the presentation, we noticed that the participants, although all living in Ouro Preto, had different backgrounds and realities: employees of the 25ª SRE-OP, students on the Teaching Technician course, students on the Mathematics degree course, elementary school teachers, nursery school teachers and Libras interpreters.

We presented the course timetable and the proposal that at the last meeting everyone would present proposals created individually or in small groups. We also explained that, at the end of each meeting, participants would complete an evaluation in which they could write down their opinions (positive and negative aspects) and suggestions for future meetings.

We then gave each participant a notebook, the Informed Consent Form (ICF) (Appendix 2, p.133) and a questionnaire (Appendix 1, p.131) to get to know them better, with questions related to their academic background, professional experience, inclusive education and, above all, their expectations of the course.

Everyone seemed interested and willing. When they arrived at the meeting place, the desks were already organized into groups. In this meeting, we had 15 participants. The seating was not predetermined; everyone could choose where they wanted to sit, forming groups:

- Group 1: Mel, Tina and Lara;
- Group 2: Simo, José, Luna and Marta;
- Group 3: Dora, Célia, Gina and Cida;
- Group 4: Livia, Rosi, Nora and Bela.

We began with an activity in which everyone had to remain silent, including us. In it, the participants had to manipulate different packaging (boxes, cans, medicine containers, tomato paste, toothpaste, etc.) and carry out tasks projected on the board, without speaking. **Activity:**

Slide 1 - LOOK AT THE SHAPES IN FRONT OF YOU

Slide 2 - DRAW WHAT YOU CAN OBSERVE

Slide 3 - WHAT GEOMETRIC SHAPES CAN YOU VISUALIZE?

Slide 4 - SHOW TO A COLLEAGUE AND ASK THEM TO EVALUATE YOUR DRAWING

---

[23] Teaching mathematics to deaf students in mainstream classes: proposals for teacher training, available at http://www.ppgedmat.ufop.br/

[24] Several parts of the chapter have been written in the first person plural. Although one of the researchers - Ana Cristina - was not present at the meetings, she took an active part in their planning, as we met weekly to plan and discuss not only the activities to be developed by the group, but also our expectations and desires in relation to the work.

[25] As we mentioned earlier, the purpose of this research was to explore inclusive practices related to teaching mathematics to deaf and blind students. One of the meetings was dedicated solely to inclusive practices related to teaching mathematics to blind people.

Slide 5 - REFLECTION

During the activity, most of the participants complied with the rules. Celia didn't complete all the tasks, as she found it difficult to recognize the mathematical concepts. Luna recognized the mathematical concepts, but couldn't associate the instruction with the way to observe. That's why we had to signal with our hands to make her understand. Everyone was watching us the whole time, as we made a positive sign to check if we could continue, and they stayed attentive to respond and follow the activity.

Only Lara (interpreter) signaled in Libras to Tina that she had a little knowledge in this area. However, in order not to give them an advantage over the other participants, we asked them to just mime using gestures.

At the end of the activity, everyone was able to express themselves freely again. We then proposed some questions for reflection:

Reflection

1)What did you think of the activity?
2) Did you understand the instructions during the activity?
3) Was it easy?

In group 1, Tina said that the activity couldn't be done for a deaf student because they wouldn't be able to read the instructions. We explained to group 1 that the text was according to the language of the participants and if we only had deaf people in the room, our text would be different.

In group 2, Luna said that it was distressing not to speak, but that everyone managed to understand each other. She also said that she thought she managed to communicate, as it was an easy task. José stressed that interacting with the group was important and gave him confidence.

In group 3, the participants didn't understand all the procedures, but they managed to communicate. Cida said it was difficult, but not impossible.

In group 4, Nora mentioned that she found it very difficult not to speak and said that they considered the activity important because they have a deaf coworker. This was one of their reasons for taking the course. The group also highlighted the importance of seeking out different forms of communication.

In the end, everyone seemed to understand the importance of putting themselves in the other person's shoes and thinking about optional forms of communication and different ways of exploring certain mathematical content. Mel mentioned the specialized material available in schools that most people never use because they don't know how to do it. Tina spoke about the lack of collaboration between school professionals and that the student is not one teacher's or another's; he or she is the school's. Lívia concluded by saying that Inclusive Education currently has many theories, but few professionals make the effort to put them into practice.

The participants, especially Lara and José, commented that the researcher's experience[26] with deaf students was fundamental in this process, since they felt a certain confidence in the way the tasks were carried out.

We then asked them to evaluate the first meeting, telling us if they had enjoyed it, if anything should be changed and if they had any suggestions for future meetings. To make them feel more comfortable, we told them that they didn't need to identify themselves[27]

In the evaluations, the participants again stressed the importance of putting themselves in the other person's shoes, praised the methodologies and gave suggestions for future meetings. They wrote that they liked the approach, which introduced knowledge through practice. Several adjectives were used to define the meeting: dynamic, great, rewarding, productive, creative and nice. Two participants wrote that they were looking forward to being able to help people with disabilities, and one participant wrote that we changed his whole concept of teaching deaf students. In terms of suggestions, participants said they would like to know how to work with Libras in the Early Childhood Education

---

[26] Viviane Cristiane Costa.

[27] This was a choice that had consequences for the research. We favored the free expression of the participants - but made it difficult to identify the authors of the ideas for the research. However, we prioritized the course and the ongoing training process.

classroom and that we should carry out more practical activities in this context.

4.2. Meeting on 28/11/15

Moments before the meeting, we had a setback. We arrived at the venue at 7.30am, but the janitor didn't show up. Then, luckily, Júlia said she knew where he lived and went to pick him up. Even so, he didn't arrive until 8:50. The school management didn't tell the janitor that he had to open the door that Saturday. Juca, José, Luna and Marta helped set up the *laptop* and *datashow*. We were able to start at 9:30.

That day, the tables were organized in groups facing the board. Each participant chose where to sit. The following groups were formed:

- Group l: Lany, Gina, Sara and Célia;
- Group 2: Simo, José, Luna and Marta;
- Group 3: Lara and Mara
- Group 4: Bela, Dara, Ñora and Leina;
- Group 5: Zeca, Júlia, Rosi and Lívia.

Noticing the presence of new members, we distributed the ICF and the initial questionnaire to them and briefly explained the dynamics of the course and its relationship to the research. Five more participants showed up: Júlia, Leina, Lany, Zeca and Dara. Tina and Mel were absent. This brought the total number of participants to 18.

So we started with a practical activity: building a Christmas star. The aim was to show the participants that we can learn mathematics (in this case, geometry concepts such as squares and their elements, flat and spatial shapes, etc.) through folding, even without speaking. We hoped they would see how simple strategies, if done well, can be carried out by all students and provide good learning opportunities.

Figure 3 - Activity: Christmas Star
Source: https://www.youtube.com/watch?v=zRfaCwuvlIQ

Everyone received two square-shaped figures and an earplug[28] . Without talking, they had to build a Christmas star by watching a video and our instructions in gestures. We also explained that, while the video was being shown, there would be breaks during which questions would be written on the blackboard for them to solve. We carried out the activity together with the participants, step by step, pausing and following along as we saw that they were able to make progress.

[28] Hearing protectors used as safety equipment (called *Push-to-Fit-Earplugs* by 3M).

Figure 4 - Start of video and first pause
Source: https://www.youtube.com/watch?v=zRfaCwuvlIQ

After stopping the video, we recorded on the board: 1) What shape is the figure?

Some participants remained unresponsive; others were eager to respond, but didn't know how. Julia even made a gesture with her fingers, but it looked like a rectangle, not a square. José wrote in his notebook that the figure was a square. So we wrote down the second question: 2) Why?

The participants began to use writing to communicate, but Júlia picked up the square on her desk and showed that its sides were the same size. Mara represented the equal sign in Libras, pointing to the sides, and Júlia added with gestures that the two pairs of sides were parallel. The participants were only able to remember the sides to explain what a square was. So we wrote it on the board:

Definition:

Square:

4 Vertices

4 congruent sides and 4 congruent angles ^ same measure (size)

To check that everyone had understood that it was a square, we made a thumbs-up sign with one of our hands. Some nodded positively and others made the same sign with the thumb of one hand. Following the video, we paused for a second time. We checked that everyone had made the indicated folds by raising the thumb of one hand in the affirmative sign.

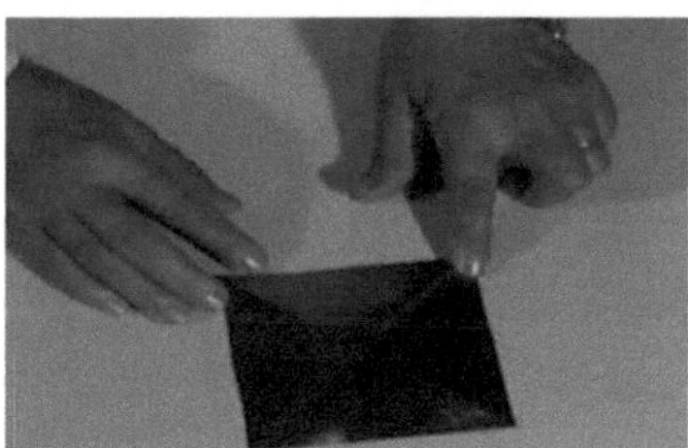

Figure 5 - Second pause
Source: https://www.youtube.com/watch?v=zRfaCwuvlIQ

Then we write another question on the blackboard: 3) What are the folds called?

We took the square shape with the folds indicated and pointed to the question. Júlia, Marta, Luna, Juca and José answered by writing the word diagonal in their blocks. The others just watched, looking unsure of how to answer. We wrote the word diagonal on the board and recorded another question: 4) Why?

At this point, everyone just watched, seemingly not knowing how to respond. So we wrote it on the board:

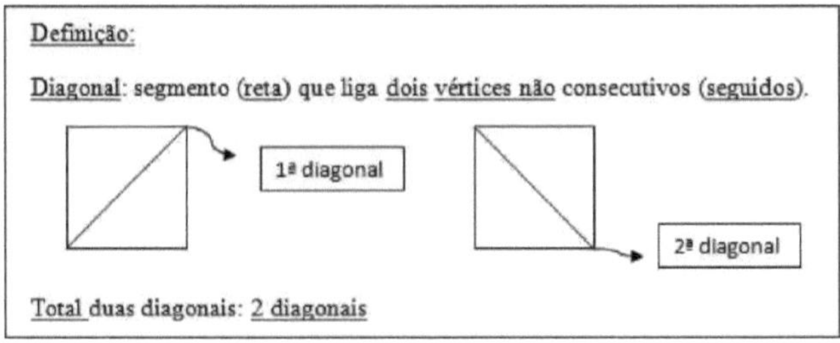
Definição:

Diagonal: segmento (reta) que liga dois vértices não consecutivos (seguidos).

1ª diagonal

2ª diagonal

Total duas diagonais: 2 diagonais

After the definition, we made the plus sign with the thumb of one hand and the participants responded in the same way. We then continued with a few more minutes of the video and stopped the show again with the following question: 5) What are the folds called now?

Figure 6 - Third pause
Source: https://www.youtube.com/watch?v=zRfaCwuvlIQ

No one could answer. We pointed out the folds and showed that each one divided the shape into two parts of the same size. We also drew pictures on the board in an attempt to represent the situation in the same way used in many math textbooks. However, none of the participants were able to name the folds.

Drawing the folds on the board:

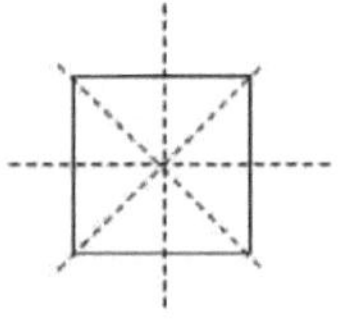

We wrote on the board, next to the previous figure: axes of symmetry. Some participants seemed to remember what it meant (group 2 and group 5) and others said they didn't remember or didn't know. We continued with the video and asked another question: 6) How or how much to cut the square?

Figure 7 - Fourth pause
Source: https://www.youtube.com/watch?v=zRfaCwuvlIQ

José and Julia tried to answer the question by explaining with their fingers and indicating on the square the size that should be cut. Célia, Leina and Lívia, however, quickly cut to the center of the square without noticing that there was a measurement. When we realized that some participants hadn't understood how much to cut, we drew a picture on the board with white and yellow chalk. We also cut the square we had in our hands slowly, trying to show everyone how to do it. We then moved from table to table, observing how each participant was doing it and helping with gestures whenever necessary.

Drawing to indicate the size of the cut in the quadrant:

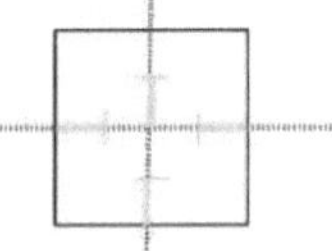

Half the side of each smaller square.

After we presented the drawing on the board and the cuts in the square, Celia, Leina and Livia - who had cut in a different way - called out to us with gestures showing a certain amount of distress. We smiled at them and handed them new pieces of paper to do the task again. We waited until the participants had reached the same stage of completing the task. After cutting out the square, everyone followed the instructions in the video and folded it again. We paused again to check that the participants were keeping up.

Figure 8 - Fifth pause
Source: https://www.youtube.com/watch?v=zRfaCwuvlIQ

There were no doubts when it came to folding. Everyone folded quickly and waited for the next step. We continued with the video. Now the participants just had to glue the edges of the squares facing inwards as shown in the video. The video was stopped again.

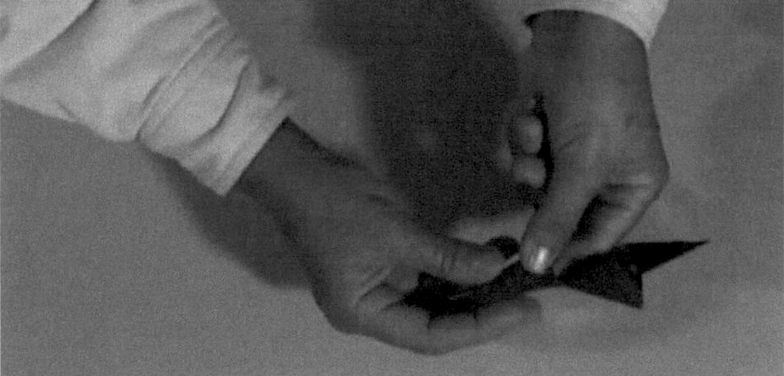

Figure 9 - Sixth pause
Source: https ://www.youtube. com/watch?v=zRfaCwuvlIQ

At this point, we went through the groups to check their collages. José, Célia, Luna and Sara didn't watch the video carefully and glued the two edges directly to the bottom of the picture without leaving any space for their fingers, according to the previous picture. With our help, some squares were recovered and glued correctly; others had to be replaced due to excess glue. After checking that everyone had managed to complete this stage, we continued with the video.

Figure 10 - Construction of the first part of the star, field notes, 28/11/2015

At this stage, the video shows us building another star in the same way using the other square that was handed out at the start of the activity. Everyone watched the video, but as they didn't seem to understand what was happening, we wrote on *the* blackboard: "*2nd square equals 1º* ." So the participants started working with the square according to the instructions and didn't even look at the video anymore.

After a while, we checked that they had built two stars according to our instructions. Once this was done, we continued the video and went to the part that showed what had to be done with the two stars. The video indicated that the stars should be glued together so that their sides were not touching. Everyone seemed afraid and more attentive. They didn't want to make any mistakes, because if they did, they would have to start all over again. So we glued together the stars we had in our hands. When José and Leina glued their stars together, they pushed too hard and ended up distorting what they were building. We showed them, using gestures, that we had to wait a while for the glue to dry. Then we had to apply the glue, squeeze lightly and place the construction on the table. In this way, the participants understood that it was more a question of skill than strength.

While the stars were drying on the table, we wrote another question on the board: 7) What happened (happen) to the initial shape (dimension)?

The participants watched and didn't want to venture an answer. To help them visualize the question better, we took the initial square and the finished star to make a comparison. We circulated part of the question: "What happened?" written on the blackboard. Some people recognized that there had been a change. Julia pointed out with her hands that we had a smooth shape and a shape with waves. José wrote the word *"changed"* on the paper. The others just watched. Faced with these reactions, we wrote on the board: Change? Everyone nodded, some with their heads, others with the thumb of one hand. Again we wrote on the blackboard: names for this change. We waited for some kind of protest, but everyone just watched. So we wrote on the blackboard: change from flat shape to spatial shape.

With everyone still silent, we wrote the word reflection on the board and handed out a sheet with some questions:

Reflections

Think for a moment about the experiences you had during this meeting... now answer the following questions without talking to your colleagues:

1) How did you feel? Please explain.

2) Did you find this activity difficult? Why?
3) What mathematical concepts could you work on with her?
4) Could this activity be useful for the Alans? Explain your answer.
5) What difficulties did you have carrying out the activity without listening? Why?
6) Could this activity be carried out in a regular class with hearing students? Explain your answer.
7) Did you think it would make it easier for a deaf student to learn? Why?

Everyone remained silent to answer the questions. They didn't show any discomfort despite spending more than an hour and a half wearing earplugs and not speaking. Still in silence, we collected the leaves as people raised one of their arms to indicate that they had finished.

When everyone had finished, we told them they could remove their earplugs and talk again. We also mentioned that we hadn't spoken for almost two hours. So we wanted to know how they were feeling after the activity. Simo said: *"I felt reflective, because we didn't speak or listen for almost two hours. And deaf people are always like that"* (meeting, 28/11/15).

We then asked them if it had been easy or difficult and what they thought of the experience. Rosi said: *"It was very difficult. I was annoyed because the teacher talks all the time, right?"* (meeting, 28/11/15)

We continued the discussion by asking if this activity could be useful for a classroom? Again, Rosi said: *"This activity could be useful for hearing students too. I think it gets the kids' attention. And by following this step by step, they become more attentive"* (meeting, 28/11/15).

And Zeca added: *"A hearing student can understand how a deaf student learns"* (meeting, 28/11/15).

Mara also spoke out:

> *I saw that the two of you were interacting in written form, right? Deaf people don't understand writing very well, sometimes they can't even draw their own letters. It's up to the person who's supporting him to teach him to hold his hand, to make strokes in a slightly different way to the way we do. So written communication is difficult, because it's part of orality. The type of inclusion of deaf students is outdated, because they are totally dependent on the interpreter or the person who supports them. The teacher can't always talk to him; the teacher confuses the good morning sign with the good afternoon sign. And the students in the class don't know how to interact with this deaf student either, so he ends up being left on the sidelines. There is no Inclusion or Inclusive School. Because inclusive school is different from inclusion. Inclusion is supported by law. Look, the school is obliged to receive, but it's not obliged to be inclusive, right? They're two different things* (encontró, 28/11/15).

As Mara looked at us, we understood that it was a question. She needed confirmation or clarification. We replied that, theoretically, based on our reading and studies of the legislation in force, the school is obliged to be inclusive. But before we could finish, we were interrupted. Mara said that the law stated that the school was obliged to receive the student. We completed her sentence and said: "to be inclusive too". And before we could say anything, Lívia said that, as of January 6th, attitudinal barriers will be punished. In response to this, Mara asked: *"And how can they be inclusive?"* (meeting, 28/11/15).

At this point, Júlia asked to speak. She said that she had taken part in the Symposium on Teacher Training and Profession (SIMPOED) in Ouro Preto and that she had learned that there is no such thing as an Inclusive School and that sometimes people don't have special needs and are excluded. He stressed that we have to respect differences, whatever they may be. She finished by saying that this was why she didn't believe in Inclusive Schools. Lívia added: *"It's a question of human rights"*. Júlia agreed.

Zeca spoke up:

> *Inclusive schools are being created, society is becoming inclusive, but it wasn't before. We used to exclude everyone, practically everyone who was different was excluded. For some time now, this has been changing and is being reflected in the school. So the teacher has to learn, the colleague has to learn, the maid, the janitor. Everyone has to learn to accept the different within the school* (meeting, 28/11/15).

At this point, Mara, who wasn't at the first meeting, said that Inclusive Education is often just talked about and not practiced. Zeca argued that people are obliged to accept this change and that, as education professionals change their perspective, everyone else will change. However, Mara insisted that Inclusive Education does not happen in the classroom; there is no action. She asked how the teacher could deal with this. Meanwhile, the other participants just watched. At this point, Livia said:

> *From the moment you see the other person in a different way, you don't have to discuss whether the school is inclusive or not; whether you are prepared or not prepared. You just have to see the other person in a different light; in a way that you expect the other person to look at you. So we have to work with this difference. It's a question of human rights. You have to look at the other person in a different way. Which isn't happening. People aren't taking the time in the classroom, above all, to look at others. They look at themselves very superficially. So, I don't think this question of saying a lot of little words really fits. I think you have to act. So, if I work in this school here, and you say that this school isn't inclusive... So, what's my role? Everyone has to do their part* (encontró, 28/11/15).

Everyone was silent. We then asked them if the activity they had just done could be done by students who are not deaf. Zeca said that the activity had been proposed in a different way with various stimuli: concrete material, tactile material, images, mimes, writing on the board and his classmates too. Júlia said that she had already used some of the things used in this activity (highlighting words, writing in different colors and folding constructions) in her classes with students who are not deaf and that she thought it was fantastic to use concrete materials for deaf students. He concluded that this activity could perfectly well be used in a classroom with both deaf and hearing students, and could also be offered only to hearing students. The other participants remained silent. After Júlia's speech, we took a 15-minute coffee break.

We then told her that we were going to watch a video of an interview with a deaf student called Camila[29] , a high school student. The video would only have subtitles, no audio, as the interviewee only communicated in Libras. Therefore, the questions were asked in Libras with the help of the interpreter who accompanies the student in her school routine. The transcript of the interview can be found in Annex 1 (p.152).

During the showing of the video, everyone remained silent and watched attentively. At the end, we commented on the context in which Camila lives and how our first contact with this student was:

> *Camila is in the 2nd year of high school.° She cried during math lessons. I thought I could work with her on high school content, in this case P.A. and P.G.[30] As I had already taught other deaf students, I decided to talk to her using signs and writing (in the same way as we did in the activity) to check her learning level. I discovered that, even though she was in high school, she couldn't add up. In view of this, supported by Inclusive Education legislation, I made modifications and adaptations to the Mathematics content that would be passed on to Camila. I came to the conclusion at that moment that, if she couldn't add up, P.A. and P.G. wouldn't make sense. So I thought I'd give her the four basic operations first, which could be useful in the future* (Field diary, meeting, 28/11/15).

After describing the context, we followed our script. We tried to start reflecting on the interview, but were interrupted by Zeca, which triggered the following discussion:

Zeca: *The class was studying P.A. and P.G. And did you work on the fourth operations with them in class? Did you teach them separately? Or did you take her out of the classroom?*

Viviane: *Yes. I didn't leave the room with her. She sits in the corner and the interpreter sits next to her. And that's the important thing. Who was my right arm, left arm, everything? The interpreter.*

Luna: Did *you plan another lesson for her?*

Viviane: *Yes, another lesson, another test...*

Luna: Would *you do another test for her?*

Viviane: I *do it because she's my student. It's geared towards her context. People sometimes think that*

---

[29] Fictitious name.

[30] Arithmetic progression and geometric progression, mathematical content taught at the beginning of the 2nd year of secondary school.°

inclusion means giving the same things and stimulating in the same way. I realize that she feels much more included now than when she was studying the same subjects as the other students.
Zeca: *Can she keep up with the class?*
Viviane: *Unfortunately, no. Mathematics is a sequence and she still doesn't understand certain concepts because she has difficulty with others. Abstraction, for example, will take a while.*
Júlia*: What bothered me the most was that nobody did anything to get her out. It went on and on...*
Luna: *It's up to the performer, you see.*
Rosi: *And the student is the teacher's, okay?*
Julia*: The interpreter is a bridge between you and the student.*
Rosi*: You've reached a point where... "she has to get out of the way. It was your stance. It was your awareness.*
Luna*: It's the teacher who has to correct the tests, right?*
Viviane: *Yes, in my case, I only do it myself.*
Lívia*: **There's** a phrase we hear from some people that goes like this: "He's an inclusive student, people, don't worry about it". And he goes on. That's why they get to secondary school without knowing how to read and write. And he doesn't know math.*
Julia*: That's a lack of responsibility on the part of the teachers! That's a crime!*
Luna*: How does it work? You give her a different test, she's learned operations and the others are in P.G. And they're all in the 2nd⁰ year of high school. Is that it? She goes from 2º year to 3 ?º*
Viviane: *Yes, the law supports it. She has the right to have the curriculum adapted.*
Julia*: She evaluates herself.*
Zeca*: But then there's the teacher's responsibility.*
Julia*: **I'm** sorry for my fellow pedagogues, but they also have to fulfill their role. The interpreter is not the class teacher. So they don't have to play the role of...*
Lara*: There are times when we feel... distressed. Because you want to help and you don't know how. Really, we get very distressed. And we end up doing it even though we know it's wrong.*
Viviane: *Look, I understand you. And that's why we're here today to look for ways to work together. I know little about Libras. What I do know I've learned from the deaf students I've had and the interpreters I've worked with. So, when we have an interpreter in the classroom, we have to combine our knowledge. I know math and the interpreter knows Libras. So let's work together to benefit the student. We have to establish a relationship of trust. When we interact with the interpreter, the deaf student feels more confident to learn and the interpreter feels valued.*

Everyone was silent. Then we emphasized the importance of everyone working together. We added that today's math teacher needs to move away from the traditional school in which only one professional is responsible for learning in the classroom. We must learn to share the attention with a support professional. We also emphasize that everyone in the classroom must be involved in this process. Zeca added that, from this point of view, students with disabilities came to the classroom to humanize the teacher and their colleagues. **Luna:** *But she said in the video that the boys didn't like talking to deaf people. Sometimes it's not even that they don't like it. Sometimes we don't know how to communicate because I don't know Libras. I'm terrible at mime... I don't know. I'm trying to learn, but there are people who can't... a student, for example, can't. Of course they have to try, but sometimes they're afraid. People are afraid because they don't know how to do it.*
José: *But people should make an effort, she said she wanted others to just say "hi", seeing that she's "there" helps* (meeting, 28/11/15).

We agreed with Luna and José. The others remained silent, seeming to agree with what had been said. We then showed *slides* exploring some concepts related to Libras.

The participants just looked at the explanations on the *slides*. Livia and Rosi tried to copy what was written. Júlia took a picture. When we showed the numbers and the alphabet in Libras, we paused and all the participants tried to reproduce each symbol they saw. We emphasized that mathematical symbols are not standardized in Libras and are usually negotiated between teacher and student. Each deaf person can represent the mathematical symbols according to their local culture. We illustrate our explanations with situations that occurred in the daily classroom life of one of the researchers as a math teacher of deaf students. The participants did not ask any questions. During the explanations, all the participants were very attentive and interested.

To end the meeting, we presented a proposal for homework: reading the text "Talking to the mothers" (available at https://novaescola.org.br/conteudo/1611/falar-com- as-maos) for next week.

We then asked the participants to record their evaluation of the meeting on a sheet of paper: what was good, what was not useful and what they would like to see covered in future meetings. Once again, the evaluations did not need to be identified. Participants were dismissed as they finished the evaluation. When they left the room, they made a point of hugging us and thanking us for the meeting.

The responses to the evaluations of the first two meetings suggest that the participants enjoyed the proposed activities, especially those in which they had to communicate without speaking. They wrote that they found communication difficult, but reported that they were surprised to learn that a teacher can and should use other means of communication besides speech. One participant wrote that he believed that only deaf people could communicate in Libras and that, by taking part in the meeting, he learned things he didn't know. They asked for more information about Libras, practical activities and for the script of the proposed activities to be sent by *e-mail.* They pointed out that changing the methodology through visualization, concrete material, step-by-step construction, exploration of mathematical content and student participation in a process based on other stimuli can improve their understanding of mathematics (for more details, see Appendix 4, p. 140).

4.3. Meeting on 05/12/15

At this meeting, four groups were formed:

- Group 1: Celia, Bela, Dora and Cida;
- Group 2: Marta, Luna, José and Simo;
- Group 3: Mara, Zeca and Lara;
- Group 4: Júlia, Rosi, Lívia and Dara.

Figure 11- Arrangement of three groups of participants, field notes, 05/12/15.

At the beginning of the meeting, mention was made of the final paper that was to be produced and presented on the last day of the course. Some of the guidelines were given in writing and commented on (see Appendix 5, p.143). Everyone listened attentively. Once again, we noticed the variation in the number of participants. In this meeting, we had the collaboration of15.

After clarifying the final assignment, we started discussing the homework. The proposal was to read the text "Talking with your hands". We asked what the text was about. Only José replied that it was about experiences and about a teacher in Bahia who faced difficulties in teaching. We asked what else we could see in the text. José replied: *"No, it was also about a geography teacher"* (meeting, 05/12/15).

At this point, Livia said that the most important thing about the text was that it showed that it was possible for the teacher to "*give"* a lesson to everyone without the need to plan differently for the students in the same class. She added: "It *often happens that the teacher prepares two lessons and, in the end, ends up "giving" none"* (meeting, 05/12/15). Júlia said that every reading we do, even if we know the text, always brings something new. She said that the text gave her other ideas and she realized that the teacher doesn't have to lower the level of difficulty of the lesson just because the student is deaf.

José spoke up again, saying:

> the *biggest difficulty described in this text was getting Libras into the context of the whole school so that everyone could interact with the deaf person. The key idea of the text is to say that there are ways to include this student; not in terms of content, but to include them socially* (meeting, 05/12/15).

We asked if anyone else would like to talk about the text. Everyone remained silent. Then, to conclude the discussion, we showed *slides* with excerpts from the text, highlighting what we thought was most relevant.

During the slide show, everyone watched attentively. So we asked who would be the protagonist of this process (inclusion). As no one spoke up, we added: *"As soon as the teacher gets involved and mobilizes other people, it's easier to start working. Now, if it doesn't come from anyone, we won't get anywhere..."* (Field diary, 05/12/15).

At that moment, Rosi said:

> *I also think, Viviane, that teachers have to feel open to learning. We see that a lot of teachers, although it may be fear or insecurity, are not very open. So I don't think the teacher can be resistant to learning* (meeting, 05/12/15).

We continued reading. José said that when the teacher of the text left the blackboard and used more visible methods, even the non-disabled students were able to learn better.

To conclude the discussion, we warn that teachers need to face up to this challenge. We emphasized that all students need to learn basic math content in order to live independently in society, regardless of whether they are deaf or not. We then asked if anyone had anything else to add. Júlia said that the deaf student's challenge was not just limited to the classroom. In addition to the school context, the family context had a big influence. She justified this by saying that, in some cases, the family doesn't accept the deaf person either. Júlia added that she really enjoyed the discussion, as she also believes that mathematics can develop an individual's autonomy. As no one spoke, we took a 15-minute coffee break.

After the coffee break, we began the second part of the meeting. Our intention was to provide the group with an activity that addressed the concept of equivalent fractions. At the suggestion of the researcher/advisor, we developed an activity using wooden fraction cubes, as shown in Figure 12:

Figure 12 - Material used in the equivalent fragments activity
Source: http://www.casadaeducacao.com.br/. Accessed on: Feb. 9, 2016.

Beforehand, we placed the wooden pieces contained in a box of fragile cubes on the tables of each group. In addition to the groups, we also had the same handles as those placed on the tables. The task was to visually understand the concept of equivalent fragments and make representations of them. The aim of the proposal was for someone else to recognize that several fragües can represent the same quantity.

We explained that everyone would have to put on their earplugs once again and that we would be doing another activity in which we couldn't use our voices, only gestures. We started by writing on the board: Equivalent Fragments. Before we gave any instructions, some people were already manipulating the material on the table. They began to arrange the shapes side by side, matching their heights. Then we raised one of our hands so that everyone would pay attention to us, starting the activity.

Next, we took one of the handles laid out on a table to represent an integer. We chose a red handle in the shape of a parallelepiped and showed it to everyone. Using gestures, we asked everyone to find a handle like it on their table.

Figure 13 - Participant carrying out the activity, encontró, 05/12/15

The whole activity took place with gestures and words written on the board. We led the process step by step and tried to accompany all the participants. At the end, we told the participants that they could remove their earplugs and speak normally. We asked them what they thought of the experiment and whether they understood what equivalent fractions are. The members of group 1 said yes. They found the approach very interesting and understood the content better the way it was presented. Cida said: *"Today, I saw sense in studying equivalent fractions"* (reflection, 05/12/15).

The members of group 2 also found it interesting, but pointed out that perhaps it would be more interesting to use the fraction disks for this activity. They argued that the disks would make the fractions more visible, reducing the risk of doubt, since the fraction cubes have many different handles which can confuse the learning process. The members of group 3 agreed with the opinion of group 2. Similarly, the members of group 4 agreed with the other groups and added that it is important to learn about other types of materials in order to offer other ways of learning to students.

Afterwards, we asked everyone to evaluate the meeting by leaving their reflections on six questions written on a sheet of paper that would be given to them. **Reflections**

Think for a moment about the experiences you had during this meeting... now, answer the following questions using the attached sheet:

1) How did you feel? Please explain.
2) What did you think of working with fractions today? Please explain.
3) Could this activity be useful for Germans or only for the deaf? Why?
4) What difficulties did you face when carrying out the activity? Why?
5) In your opinion, what is the role of the teacher who teaches mathematics in a heterogeneous class with hearing, deaf and other students? Explain your answer.
6) What does a teacher need to teach math to children and teenagers in order to do a good job, even with varied classes?

While people were responding to the reflections, we took the opportunity to give a reminder about the final work.

Afterwards, we read the aforementioned reflections. In the first question, we found that six participants felt good during the activity and thought it was a great experience. One person wrote that they felt uncomfortable because the groups formed among the participants were very homogeneous. Five participants wrote that they felt distressed at having to remain silent and communicate only by gesture. Another wrote that he felt alert, because he had to be attentive at all times so as not to miss any stage of the activity. Only two participants mentioned the importance of the methodology in improving learning:

> *Being deaf doesn't make me incapable of learning, as long as the guidance is targeted, clear and objective* (reflection, meeting, 05/12/15).
>
> *I was surprised by the way the lesson was given. And how the search for change in the classroom can improve student performance* (reflection, meeting 05/12/15).

Everyone liked the fractions content worked on at the meeting. We obtained the following classification:

Opinion	№ of participants
Great	4
Very good	4
Very interesting	3
Very clear	2
Wonderful	1
Challenging	1

Table 1 - Participants' opinions on the content of working with fractions Source: Field notes, meeting, 05/12/15.

The justifications made reference to the concrete material used, highlighting the importance of diversifying methodologies. On the third question (about the usefulness of this activity in the ordinary classroom), everyone answered in the affirmative. The participants explained that this activity would make it easier for students to learn.

When it came to the difficulties faced during the activity, of the 15 participants, only three wrote that they found it difficult to communicate; in other words, to understand our gestures and communicate without speaking. The others wrote that they found it difficult to remember and associate the mathematical content with the concrete material.

In their opinions on the role of the teacher who teaches mathematics in a heterogeneous class, with hearing, deaf and other students, all the participants wrote that the teacher needs to look for new methodologies in order to better meet the needs of their students. Most of the participants did not specify what this should look like. Three participants specified that the teacher should try to teach everyone homogeneously, using the same resources and including everyone in the same activity. And one participant, in addition to sharing our ideas from the reflection on the video of Camila's interview, argued that the main thing is to stimulate learning, regardless of whether it is homogeneous or not:

> *The teacher's role is important, as he or she is the protagonist of the process. They should look for different methodologies, because students are different. The important thing is to stimulate learning, whether collectively or individually* (reflection, meeting, 05/12/15).

In the final question about what teachers need to teach math to children and adolescents in order to do a good job, even with varied classes, the participants were unanimous. They answered that teachers need new knowledge, which can be acquired through training, research or meetings with school specialists. They all stressed that new knowledge, as well as teaching methodology, must take into account the particularities of each student.

4.4. Meeting on 12/12/15

In this last meeting, the participants had to present, individually or in groups, a task or a teaching proposal for mathematical content from an inclusive perspective. They could develop a simple task, a more elaborate activity or even a teaching proposal involving several lessons. In this meeting, the tables were organized in a U-shape.

There were 13 participants:

- four students from the Mathematics degree course (Luna, Marta, José and Juca);
- four students from the Technical Teaching course (Gina, Célia, Cida and Dora);
- three representatives from the Department of Education's training department (Lívia, Rosi and Zeca);
- a math teacher (Júlia);
- a Libras interpreter (Mara).

Figure 14 - Layout of the participants at the last meeting, 12/12/2015.

We then asked who would like to start the presentation. José, a student on the Mathematics degree course, volunteered. He planned his activity individually. He proposed working with the abacus. Showing *slides* with images of various ancient and current abacuses (Mayan, Chinese, Roman, Japanese, Russian, didactic and adapted), he commented a little on their history since their beginnings in the Mesopotamian region. He mentioned that the word "abacus" means "board" and that the first models were made out of sand.

José explained that each abacus has its own peculiarities. He pointed out that the Russian abacus is the most similar to the didactic abacus used in Brazil. He then explained that didactic abacuses are the abacuses used in schools. He took several copies out of a bag and gave them to the other participants.

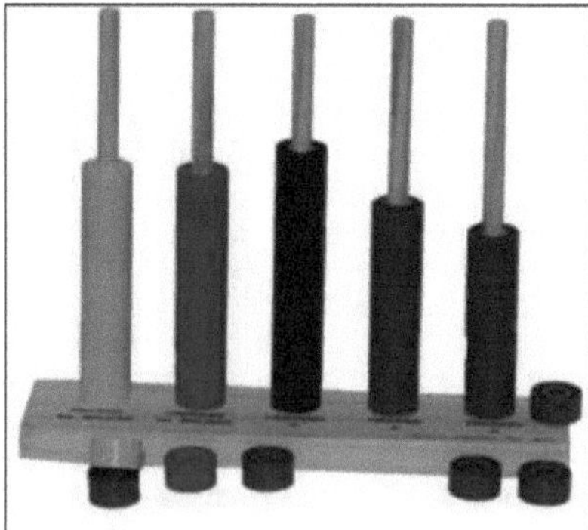

Figure 15- Didactic abacus shown by José
Source: https://www.casadoeducador.com. Accessed on: 20 Mar. 2016.

Continuing the presentation, José said that not all schools have the abacus, so he would talk more about the adapted abacus. He pointed to the *slide* and described the materials used to make the abacus. He explained that the teacher could make their own abacus and that, depending on the material used, it could be more durable or less durable (more sophisticated or less sophisticated). He then talked about the usefulness of the abacus. She said that it is useful for children who are in the process of learning addition or subtraction. You start with basic operations and then increase the level according to the student's development. He pointed out that the activities that can be carried out on the abacus are suitable for all students (deaf and hearing).

To show how useful the abacus is for teaching addition and subtraction, José explained how it works. As he had distributed several abacuses, he asked the participants to record some numerals. He then moved on to the operations. After recording the numeral 84 on the abacus, José asked how much they would like to add to it. Zeca answered 120. José explained how to proceed in detail.

After the explanation, Zeca asked how to subtract 70 from 204 on the abacus. José explained how to do it, while everyone followed along with interest. He went on to explain how to do multiplication and division on the abacus using examples. He concluded by emphasizing that the key is to understand the structure of the abacus in order to understand the operation.

We asked José how he would present the abacus to a deaf student. He promptly replied that he would pass everything on to the student's Libras interpreter, as he doesn't know how to communicate in Libras. So we asked him another question: *"Would you be able to teach this deaf student anything?"* As José remained silent, we asked: *"But couldn't what you see here in this course help?"* (Viviane, meeting, 12/12/2015) We asked José to imagine that the interpreter was absent from work that day, and asked if the deaf student would miss school. A little shyly, he replied:

> *I was going to do what I've just done here. Quite calmly, I was going to show him that 10 units equals one ten; I was going to show him that...* [took the 10 pins and put them on the units stick] *... 10 is equal to this* [took a pin and put it on the tens stick] (meeting, 12/12/2015).

We added: *"And you've learned the equal sign. What is it?"* (Field diary, meeting, 12/12/2015). To answer, the participant correctly made the sign that corresponds to equality in Libras. We complimented her by saying very well. At this point, Luna talked about the cell phone apps that exist for learning Libras. She said she couldn't remember the name, but that the teacher could speak into the cell phone and a doll would translate it into Libras. She added that these apps could be very useful when communicating with deaf people.

At this point, we asked them which resource they used most during our activities. We said they didn't know Libras, but we couldn't speak either. José replied: *"Writing, underlining the important words... being patient... having persistence"* (José, meeting, 12/12/2015). We added that one of the most important things in this situation had already been said by another participant in a previous meeting: the teacher needs to get out of their seat.

The next person to introduce herself was Mara. She brought a small bucket full of Lego pegs and the article "How to teach math with Lego pegs?"[31] (HILTON, 2015). She began her presentation by reading the first paragraph of the article: *"It's almost impossible to find someone who doesn't know what Lego is. A phenomenon in the 80s and 90s, these building blocks were practically the norm among the toys of those times"* (HILTON, 2015).

He added that Lego exists in other countries, not just Brazil, and that it is a worldwide trend. He reported that in this article the author shows the "concreteness" of Lego and gives parameters for its use in teaching mathematics. It showed that this toy can be used to teach math to deaf students.

He then showed a *slide* with the following picture:

[31] Article available at: <http://www.tudointeressante.com.br/2015/12/como-ensinar-matematica-com-pecas- lego.html>. Accessed on: 3 Mar. 2016.

Figuraló - Lego piece with 48 points displayed

Source: www.tudointeressante.com.br/2015/12/como-ensinar-matematica-com-pecas-lego.html

Mara took a Lego peg similar to the one on the *slide* and explained how to do a division using it.

> *He* [the author of the text] *says to get a handle with 48 points. It would be a whole Lego, just like the golden material that gives that square that represents the thousand. But then he says to take this peg with 48 points and then divide it into other pegs that fit there* (meeting, 12/12/2015).

Figure 17 Mara explaining the Lego peg with 48 points shown on the *slide* (meeting, 12/12/2015).

Next, Mara picked up a Lego peg with two dots. Using *slides*, she showed how to use each Lego peg to represent a different fraction:

Fração	Fração
Número Inteiro	

Figure 18 *Slide*: Lego handles representing fractions

Source: www.tudointeressante.com.br/2015/12/como-ensinar-matematica-com-pecas-lego.html

She also said that since almost every child or teenager has this toy, the student can continue practicing the material they saw at school at home in a simpler way. She added: *"It's playing while studying or studying while playing"* (Mara, meeting, 12/12/2015).

She then presented *slides* showing fraction additions using more than one Lego handle. Mara commented that she had found the approach very interesting, but that unfortunately she didn't know how to explain the additions shown on the *slides* correctly.

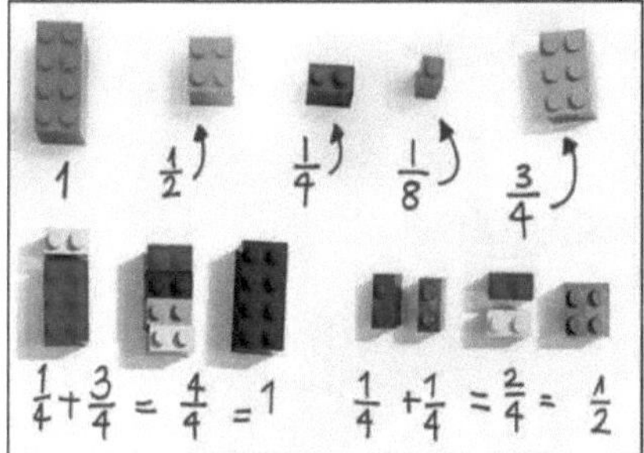

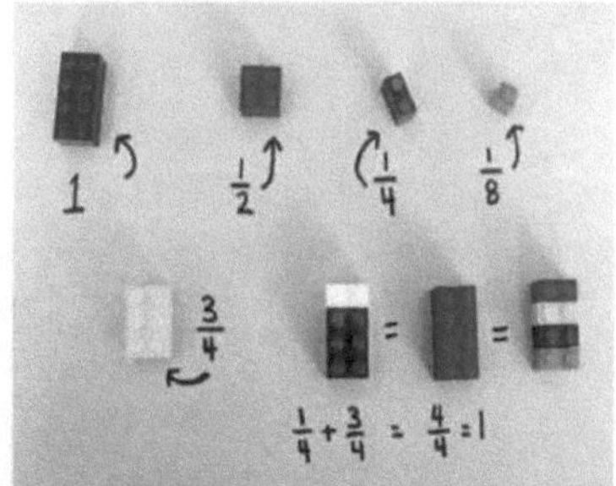

Figure 19 - *Slides*: Lego handles representing the addition of fractions
Source: www.tudointeressante.com.br/2015/12/como-ensinar-matematica-com-pecas-lego.html

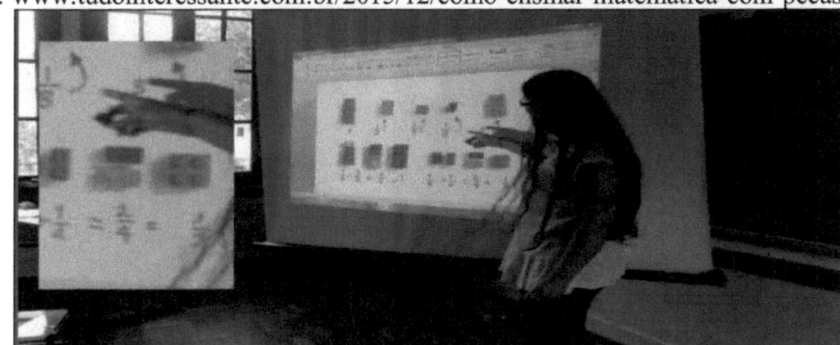

Figure 20 - Mara showing a slide with a Lego stick in her hand to represent the addition of fragments (meeting, 12/12/2015)

Other mathematical content was also covered. Mara also showed that Lego pegs can be used to represent perfect square numbers:

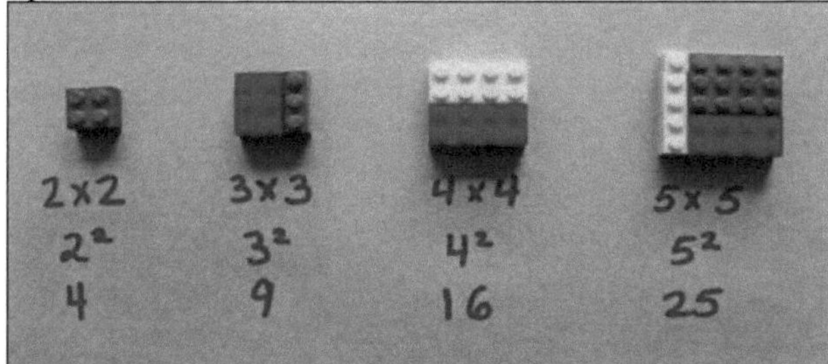

Figure 21 - *Slide*: Lego pegs representing perfect square numbers
Source: www.tudointeressante.com.br/2015/12/como-ensinar-matematica-com-pecas-lego.html

and arithmetic mean:

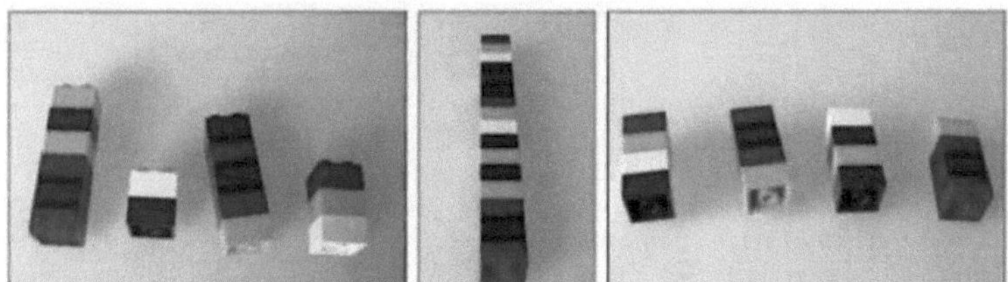

Figure 22 - *Slide*: Lego handles representing arithmetic mean calculation
Source: www.tudointeressante.com.br/2015/12/como-ensinar-matematica-com-pecas-lego.html

Figure 23 - Mara showing *slide*: calculating arithmetic mean with Lego handles (encontró, 12/12/2015)

During the presentation of the *slides*, Mara said once again that she wouldn't be able to explain the mathematical content exactly. However, she gave the participants the source of the material she had presented (www.tudointeressante.com.br). She said that, in her presentation, she focused more on teaching mathematics in the classroom, especially to deaf students. Júlia was interested and asked Mara if her work was in a resource room or in a regular classroom. Mara replied that she accompanied a deaf student in a regular classroom.

We asked her if the use of Lego was only for a deaf student or if hearing students could also use it. Mara replied that it was an activity for everyone, as the toy she brought to the presentation belonged to her son, who was not deaf.

We took a 15-minute coffee break and then continued with the presentations.

Júlia, representing her group, began by stating that they were going to propose an activity using Tangram with the aim of teaching mathematics to the deaf. This activity had been planned together with two other colleagues from the course: Lara and Bela. Unfortunately, due to a personal problem, Lara was unable to attend the presentation. So the activity was presented by Júlia with Bela's help. Julia then showed the following *slide*:

Figure 24 - First *slide* shown by Júlia's group (field notes, meeting, 12/12/2015)

While Bela handed out an A4 sheet of paper to each participant, Júlia explained that she was going to propose a practice with Tangram in two parts: in the first, she would describe what Tangram was and its history; and in the second, she would propose tasks for her colleagues:

> *Tangram is a Chinese puzzle. A puzzle made up of seven pegs. And you can assemble more than 1,700 figures. We can work with it as a game, understanding geometric shapes, fractions, proportions, creativity in the sense of art and logical reasoning* (meeting, 12/12/2015).

Figure 25 - Júlia presenting the Tangram to the other participants (meeting, 12/12/2015)

After presenting some legends about the Tangram, as well as its characteristics and the objectives of the presentation, Júlia said that everyone was going to build a Tangram. She showed two *slides*: one with the material needed and the other with a figure of a square. She asked participants took the A4 sheet of paper that Bela had distributed. With the sheet in her hands, she began: *"It's quite simple. We're going to meet the smaller part of the rectangle with one of the larger sides so that it's right at the end. Then you'll see that there will be a piece left over that we can fold and cut"* (meeting, 12/12/2015).

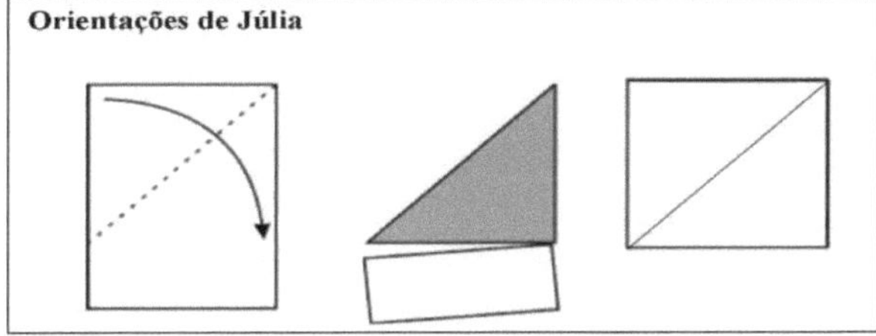

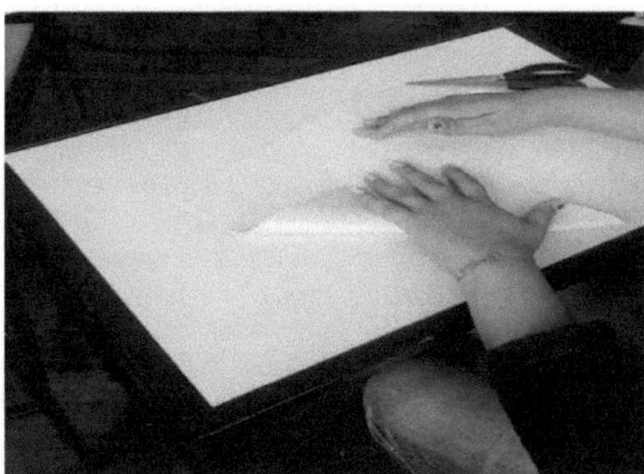

Figures 26 and 27 - Orientations presented and Dora following the orientations (meeting, 12/12/2015)

After constructing the square, Julia continued: *"Now, we're going to mark one of the diagonals with the pencil. Then we're going to cut that mark you made"* (meeting, 12/12/2015). She then showed the participants that if they cut the square in the way indicated, they would get two triangles.

Figura 28 - Júlia showing the two triangles to the participants (meeting, 12/12/2015)

Julia checked that all the participants were able to complete the activity. She asked them to take one of the triangles, fold it in half, mark it with a pencil and cut it again. He then took one of the triangles obtained from the cut and the triangle that wasn't cut, and added a few comments: *"Going back to the two original triangles, we see that they are similar and that in relation to this third one, which we have obtained, it is half of that one. So you see: similarity, proportion and geometry, all in one situation*" (Júlia, meeting, 12/12/2015).

Next, Júlia took the other triangle that hadn't been cut and asked the students to fold it in the same way as she was folding and cutting it. She used very simple language, just pointing out with her fingers how it should be charged: *"Here, there, fold there, that point at that little point*". She made a point of saying the name of each geometric figure she was constructing. As he spoke, he made sure everyone was following along.

Figura 29 - Luna doing the orientation at° 11 (met, 12/12/2015).

While Júlia was showing the procedures, she asked the others what the name of the figure they had just found was. This happened when they formed a triangle and a trapezoid (orientation no. 6), a square and a trapezoid (orientation no. 16) and then when they found a parallelogram and a triangle (orientation no. 22). However, she didn't wait for the participants to answer, instead anticipating them and answering her own questions. Having built the handles, Júlia said: *"The first challenge now is to assemble the square again"* (meeting, 12/12/2015). Luna said: *"Eita"*, and Mara quickly took the two largest triangles and put them together to form the square. Then she asked Júlia if it was right. Júlia told Mara that the square had to be built using the seven handles of the Tangram.

The participants tried to assemble the square and Júlia, seeing that they were having difficulties, said: *"You have to! Why don't we cut out a square to make these handles?"* (meeting, 12/12/2015). She waited a little longer, asked if it was ready and they all said no. As she didn't have much time, Julia said she would reveal the secret and explained how to solve the challenge.

Figure 30- Júlia asking the participants to join the two larger triangles (encontró, 12/12/2015)

Marta, José, Bela, Mara, Gina, Rosi, Zeca and Juca managed to put the square together without any difficulty. Luna managed to put it together with José's help. Lívia put it together with Júlia's help. The other participants - Dora, Célia, Dora and Nora - couldn't put it together because, while they were trying, Júlia moved on to another challenge.

Figure 31- Dora having difficulty assembling the square (meeting, 12/12/2015)

The second challenge consisted of assembling a cat with the seven handles of the Tangram. The image was projected and Júlia asked the participants to reproduce it.

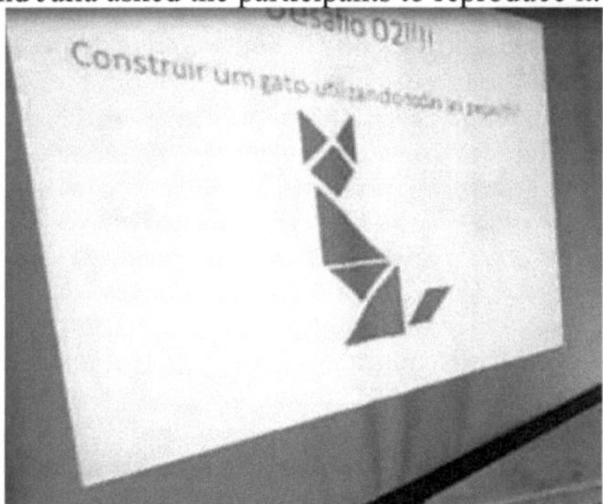

Figura 32 - *Slide* presented to build the cat (field notes, encontró, 12/12/2015)

Figura 33 - Cats built by Marta and José (meeting, 12/12/2015)

Julia then showed another *slide* showing other positions for building the cat.

Figure 34 - *Slide* shown by Júlia with other ways of assembling a cat (meeting, 12/12/2015)

After showing the cat *slides*, Júlia launched into a third challenge, but emphasized that this challenge could only be proposed to elementary school students once they were very familiar with Tangram. She then presented the solution to the challenge and proposed others.

The participants were attentive to everything:

> *We can use Tangram in a practical way: to make a title for a panel, a drawing. And not just you. The teacher doesn't have to do the activity. It's the students who have to do it! Show them, the first time, teach them, work with them. Leave it up to them. As long as math is seen in a dry way, we will always keep hearing the phrase: math is difficult, I don't like math, I can't learn math. I've heard that a lot. So, really, if it's just blah blah blah... it's not possible. You can work so hard on the theory and bring something to life* (meeting, 12/12/2015).

Finally, Júlia drew seven colored Tangrams, which she had made at home, and said:

> *The interpreter, as she* [Viviane] *said, is a help in the classroom, but he's not the one who has to take on all the responsibility. He's going to help me, be my right and left arm. The responsibility lies with the class teacher. We have to be aware that the support teacher has to support and do so* (meeting, 12/12/2015).

The next presentation was by Luna. She began by apologizing for presenting such a simple paper. She explained that she hadn't had access to the internet the day before. So she couldn't print out her work. She made a fraction domino to work with the deaf students and explained: *"On this domino, we have fraction representations: on one side, I have the fraction in... numbers? How do I say it? Is it a number?"* (meeting, 12/12/2015) Julia replied: *"It's an algorithm"* (meeting, 12/12/2015). Luna thanked her and continued presenting:

> *So, on the one hand, I have the algorithm and, on the other, we have the picture*

representation. The idea is to find the handles that fit. The cardboard I used was able to make 25 handles, it took a long time because fractions are equal parts and I did everything with a ruler. You can't give the whole design away. But there are ready-made dominoes on the internet. Here, I've done it only with rectangles, but there are ones that use a circle. You can find lots of different shapes. And this activity is so easy for a deaf student... (meeting, 12/12/2015).

José interrupted Luna to point out that the dominoes could also be used to work on equivalent fractions. Luna replied: *"And improper fractions too. I didn't just draw the pretty fractions. There are proper and improper ones. Students who are deaf or not can have access to everything* (encontro, 12/12/2015).

Figure 35 - Luna presenting the fragóes dominoes (encontró, 12/12/2015).

Figure 36 - Some handles from Luna's fragóes dominoes (meeting, 12/12/2015).

As Luna and the other participants remained silent, we commented that the idea of dominoes is very interesting and can make it easier for deaf students to understand mathematical content.

We then asked everyone to evaluate the course with another questionnaire. As the participants finished, we said goodbye and thanked them once again for their participation. Júlia, Zeca, Rosi and Lívia were sorry that this was the last meeting. Everyone was very affectionate, said goodbye and thanked us for the experience. At the end of the course, we read their evaluations (see Appendix 3, p.135).

Chapter 5

Analyzing Júlia's trajectory

> *In description, the opinions of the informants are presented as faithfully as possible, as if the data spoke for itself; in analysis, the purpose is to go beyond what has been described, breaking down the data and looking for relationships between the parts that have been broken down and, finally, in interpretation - which can be done after analysis or after description - we look for meanings in the speeches and actions in order to arrive at an understanding or explanation that goes beyond what has been described and analyzed* (MINAYO, 2011, p.80).).

Among the different paths taken by the participants, we have chosen to narrate Júlia's here. Firstly, because she is an experienced math teacher, unlike all the others; and secondly, because she was already working as a math teacher when the Inclusive School was implemented. She was also the only participant who openly stated at the first meeting that she didn't believe in the Inclusive School.

Júlia has a degree in Civil Engineering and Pedagogy. As we reported, she had been teaching mathematics in elementary school and in a technical teaching course in a public school in Ouro Preto for more than five years. Our contact with her throughout the course provided evidence that she is a respected professional who likes to express her opinion and who needs consistent arguments to be convinced of something different.

Although he missed the first meeting of the extension course, he actively participated in all the others. She brought along some students from the Technical Teaching course, for whom she taught, but she always preferred to sit in the group made up of staff from the 25ª SRE-OP. Constantly attentive, she helped the members of her group with their tasks. During reflections on the activities, she always expressed her opinion or argued about the opinions of others. In the evaluations of each meeting, she often made a point of saying thank you and signing, even though we made it clear that they didn't need to identify themselves.

On the second meeting of the course, Julia's first day, we presented the work proposal, which involved reading and discussing short texts on inclusion, but above all experiencing and constructing proposals for teaching mathematics to deaf people.

We emphasized that, as far as possible, we would work with the perspective that the proposals presented could be developed with the whole class, and not just for the students.
deaf. With this, we expressed our concept of inclusion as a broader perspective, involving classes and the school as a whole in respecting difference.

At this meeting, when answering a questionnaire, Júlia said:

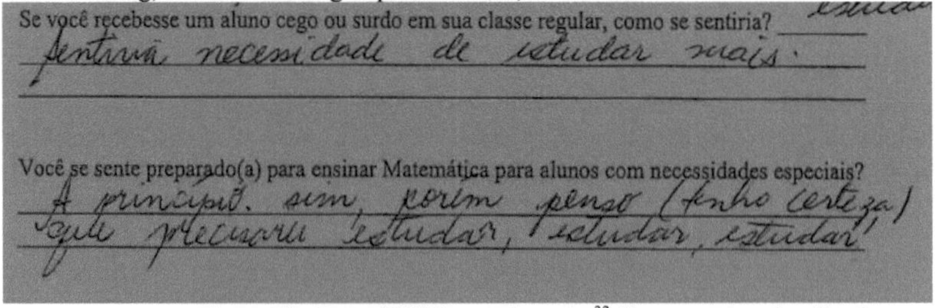

Se você recebesse um aluno cego ou surdo em sua classe regular, como se sentiria?
Sentiria necessidade de estudar mais.

Você se sente preparado(a) para ensinar Matemática para alunos com necessidades especiais?
A princípio, sim, porém penso (tenho certeza) que precisarei estudar, estudar, estudar!

Figure 37 - Part of Júlia's questionnaire[32] Source: research collection.

Our intention in proposing these questions was to provoke the participants to reflect on the inclusion of deaf students in mainstream schools.

When answering the questions, despite having more than five years' experience as a math teacher, Júlia said that she needs to study more, as she is facing a new situation[32][33] . In the same questionnaire, it is interesting to note that she explains her initiative in taking part in the extension course by her desire or need to study ways of teaching mathematics to deaf students. She also points out that the idea of collectively building inclusive teaching activities and proposals seems very good to her, because specialized material is scarce or difficult to access (expensive).

Júlia's answers confirm the considerations of Gil (2007) and Souza (2013). The authors emphasize that there is a need for greater investment in the training of math teachers who teach the deaf, so that they can collaborate in the construction of curricula aimed at Inclusive Schools, as determined by current educational legislation.

The teacher ends the questionnaire with a suggestion: *"I would like to learn how to teach fragao to a deaf student"*. In our opinion, Júlia expresses the need to develop her pedagogical knowledge of the content (SHULMAN, 1987).

In this sense, pedagogical content knowledge lies in the teacher's ability to transform his or her knowledge of content into forms of action that are pedagogically effective for student learning. It represents an amalgamation of subject matter and didactics, through which an understanding is reached of how certain subjects or problems are organized, represented and adapted to the diverse interests and abilities of students. This type of knowledge, according to Shulman (1987 *apud* CASTRO, 2009), makes it possible to differentiate between the understanding of a specialist in an area and that of an educator.

In the first task, as described above, we proposed the construction of a Christmas star. Júlia collaborated with the activity both in building the star and in exploring mathematical concepts. When we wrote questions on the board for the participants to answer without speaking, she remained attentive the whole time. She built the star by observing the instructions without using speech. She identified almost all the mathematical concepts involved. She tried to answer the questions using gestures and writing. After everyone had managed to build the star, still in silence, we wrote the word reflection on the board and handed out a sheet with some questions. Julia answered:

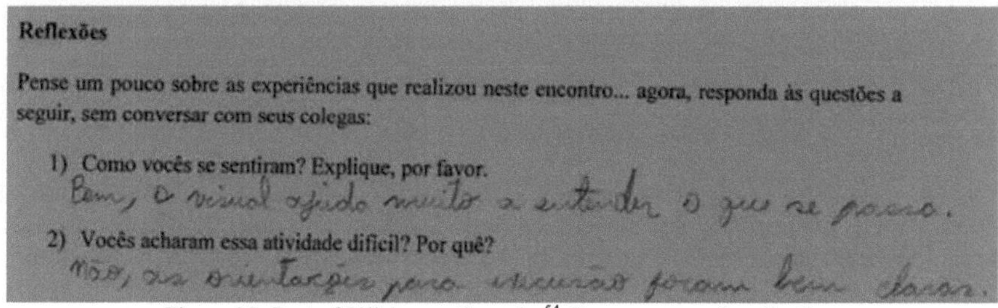

Reflexões

Pense um pouco sobre as experiências que realizou neste encontro... agora, responda às questões a seguir, sem conversar com seus colegas:

1) Como vocês se sentiram? Explique, por favor.
Bem, o visual ajuda muito a entender o que se passa.

2) Vocês acharam essa atividade difícil? Por quê?
Não, as orientações para execução foram bem claras.

Figure 38- Part of Júlia's questionnaire[54] Source: research collection.

We noticed that she felt comfortable doing the activity and that working with visualization helped her understand the content. The intention of this activity was precisely to explore mathematical concepts using visualization, because the particularity of deaf students does not restrict their vision. Like Sales (2013), we believe that visualization is important in teaching mathematics to all students, including deaf students.[34]

---

[32] If you had a blind or deaf student in your class, how would you feel? "I would feel the need to study more". Do you feel prepared to teach mathematics to students with special needs? "At first, yes, but I think (I'm sure) I'll need to study, study, study".

[33] In the same questionnaire, we asked the participants if they had ever taught students with special needs. Júlia replied: "Not yet".

[34] How did you feel? Please explain. "Well, the visual helps a lot to understand what's going on." Did you find this

> In the educational context, this issue should be reflected on when thinking about teaching mathematics, when thinking about the types of visualization needed by students. In both mathematical and other contexts, visualization is related to the ability to: create, manipulate and 'read' mental images, guiding and assisting in the formation of logical connections and demonstrations; visualize spatial, quantitative information and visually interpret the information presented to them; review and analyze steps previously taken with objects they can touch, draw and interpret (GOLDENBERG, 2010 *apud* SALES, 2013, p.70).

In the same vein, Nunes *et al.* (2011) emphasize the effectiveness of mathematics activities that explore the visual aspect through interventions designed for their specific needs. Seventy-five British deaf pupils, with an average age of six to seven, and their teachers took part in the study. The authors determined three levels of analysis: a measure of their mathematical reasoning, an assessment of their cognitive abilities and an evaluation of their working memory using individualized tests that already exist and are used in that country. Among the mathematical contents discussed were: additive composition of numbers and additive and multiplicative reasoning. In the tests applied by the authors, various materials were used, as well as pictures, in a visual presentation of the mathematical information (BORGES, 2013).

When Julia answered the fourth question, she associated the star activity with other mathematical concepts. She explained that, with this same activity, she would work on geometry (plane and spatial) and the properties of quadrilaterals such as: diagonal, angle, symmetry and internal angles (sum).

According to Shulman (1987), a teacher's professional development process can be represented by a Pedagogical Reasoning and Action Model (PRAM). According to the APRM, given a teaching text, educational objectives and a set of ideas, pedagogical reasoning and action involves a cycle through activities of understanding, transformation, instruction, evaluation, reflection and new understanding (FERNANDEZ, 2015).

Another answer that caught our attention was Julia's to the fifth question:

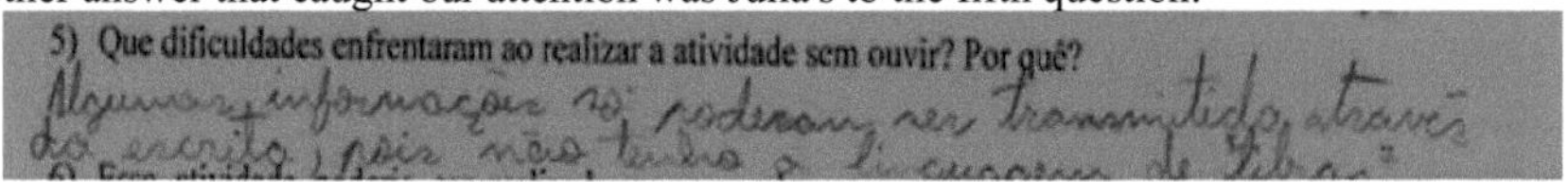
5) Que dificuldades enfrentaram ao realizar a atividade sem ouvir? Por quê?
Algumas informações só poderam ser transmitidas através do escrito, pois não tenho a linguagem de Libras.

Figure 39 - Júlia's reflections[35] (cont.) Source: research collection.

Although Júlia pointed to language as a difficulty in carrying out the activity, she had previously pointed out that the instructions were clear. In this way, we realized that when she didn't understand the content just by looking at it, she resorted to writing. However, this writing did not simply represent the act of writing.

[35] What difficulties did you have carrying out the activity without hearing? Why? "Some information could only be transmitted through writing, because I don't 'have' the Libras language."

During the star activity, starting with mathematical content (properties and elements of the square), we wrote questions on the board and highlighted the words that had a translation into Libras. Júlia realized that, even without knowing Libras, she could communicate with me and understand the mathematical content without speaking or writing for long periods. She wasn't the first to communicate in writing. She began to do so after observing a classmate.

The teacher realized that writing in a few words would make it easier for a deaf student to understand and that there are other methodologies for teaching mathematics. However, it is interesting to note that this writing was possible. During the activity, we used these aspects to write on the board. We highlighted the words that had an interpretation in Libras. Sales' (2008) studies state that the use of Libras, associated with some teaching resources, can establish a communication channel favorable to collective interaction and the appropriation of mathematical concepts.

Neves (2011) argues that when teaching mathematics to the deaf, the mathematics teacher

---

activity difficult? Why? "No, the instructions were very clear".

should structure methodologies according to three competencies: mathematical knowledge, Libras and Portuguese language. According to her:

> Our research shows that mastery of the triad of mathematical knowledge, LIBRAS and Portuguese is essential for teaching these students. The lack of mastery of one of these three areas of knowledge results in methodological obstacles, contributing to the realization of learning obstacles (NEVES, 2011, p.111).

In our research, we don't prioritize the signaling of words in Libras, but rather the structural aspects of the language in order to facilitate communication between deaf and hearing people.

During the course, we asked the participants if the star activity could only be developed with deaf students or if we could involve the whole class. Júlia said:

> *I've already used some of the strategies proposed during this activity: highlighting words, writing in different colors and constructing foldables in my classes with students who are not deaf. I thought it was fantastic to use concrete questions for deaf students. This activity could perfectly well be used in a classroom with deaf and hearing students. Or it could be proposed just for hearing students too* (Field notes, 28/11/15).

This response suggests that she sought to diversify the strategies used in her classes, corroborating what other scholars have already mentioned in their research. Oliveira (2005) used Origami as a resource in Geometry classes for the deaf with hearing teachers and showed that, as well as adding mathematical content, it can break down the communication barrier between teachers and students. As described above, Santos (2012) developed geometry activities involving the contents of symmetry and reflection. The activities were carried out by deaf and hearing students with the help of the Transtaruga microworld[35] . The results indicated that all the students were involved with the mathematical properties of the symmetry and reflection content due to the interaction of linguistic resources and physical manifestations related to the movements of the computational objects on the computer screen. Diversity is interesting for all students, including the deaf. It favors learning insofar as it allows for different ways of constructing knowledge.

Below are some of Julia's comments that have sparked our reflection.

In the first questionnaire, we asked the participants: *"Nowadays, there is a lot of talk about inclusion and inclusive schools. How do you understand these terms?"* Júlia wrote: "Inclusion: respecting differences and realizing that the world is diverse. I don't believe in inclusive schools".

Reflecting on the star activity, Júlia recognized that the activity could be carried out in a classroom with deaf and hearing students. This gives us indications of Júlia's first perceptions of inclusion and the mobilization of pedagogical knowledge:

> Pedagogical knowledge is presented as doctrines or conceptions arising from reflections on educational practice in the broad sense of the term, rational and normative reflections that lead to more or less coherent systems for representing and guiding educational activity (TARDIF, 2002, p.37).

When answering another question in the first questionnaire, Júlia said that she knew a teacher who had a deaf student, who dealt with the situation calmly because there was an interpreter in the classroom. For her, the teacher's only difficulty was communicating with the student because he didn't know Libras. These ideas lead us to think that she believed that the presence of an interpreter was the only requirement for a deaf student to be included in the classroom, because, according to Borges (2013):

> The characteristics we observed in the inclusion of deaf students in math classes lead us to assume that the teacher understands that the deaf student 'is not my student', that their problems 'are not mine', that their learning 'is not my responsibility', but rather that of the Libras interpreter (p. 173).

However, after the video of an interview recorded with Camila was shown, Julia seemed to revise her position:

Júlia*: What bothered me most was that nobody did anything to get her out of the place. It just kept going....*

---

[35] Dynamic interactive computer environment with a language and computer use adapted to the mathematical content.

Luna (student majoring in Mathematics)**:** *It's up to the interpreter, see?*
Rosi (an employee of the Education Department): *And the student is the teacher's, right?*
Julia*: The interpreter is a bridge between you and the student.*
Rosi*: You've reached a point where... "Hold on. She has to move. It was your stance. It was your awareness.*
Luna*: It's the teacher who has to correct the tests, right?*
Viviane: *Yes, in my case, I only do it myself.*
Lívia: (an employee of the Department of Education): *A phrase we hear from some people is: "He's an inclusive student, people. Don't get upset". And he gets through. That's why they get to secondary school without knowing how to read and write. And he doesn't know math.*
Julia*: That's a lack of responsibility on the part of the teachers! That's a crime!*
Luna*: How does it work? You give her a different test, she's learned operations and the others are in P.G. And they're all in the 2nd⁰ year of high school. Is that it? She goes from grade 2º to grade 3 ?º*
Viviane: *Yes, the law supports it. She has the right to have the curriculum adapted.*
Julia*: She evaluates herself.*
Zeca (an official from the Education Department): *But then there's the teacher's responsibility.*
Julia*: **I'm** sorry for my fellow teachers, but they also have to fulfill their role. The interpreter is not the class teacher. So they don't have to play the role of.*

Júlia seems to be rethinking her view of the roles of the interpreter and the teacher: *The interpreter is a bridge between you and the student". "What bothered me most was that nobody did anything to get her out of her seat. It just went on and on..."* And when she said: *"I'm sorry to my fellow pedagogues, but they also have to fulfill their role. The interpreter is not the class teacher. So they don't have to play the role of a teacher"* (Júlia, field notes).

In addition, after watching the video with Camila's interview, he evaluated the behavior of the teachers: *"This is a lack of responsibility on the part of the teachers! That's a crime!"* And, from the context, she even reflects on Camila's assessment: *"She is assessed by herself"* (Júlia, field notes).

As we mentioned earlier, the teacher said twice that inclusion was a matter of respecting differences and that she didn't believe in inclusive schools.
When talking about the teacher's relationship with the deaf student, she said that it went smoothly. Camila's experience throughout her student life does not seem to us to be any different from the situation of the teacher reported by Júlia. In our opinion, the situation has just changed narrator. However, when we watched the interview, she didn't signal it in the same way. According to Shulman (1996), stories that start from experience can be transformed into cases through narration. And the plot of each case involves a context, an intervention that is not complete; it makes it possible to create situations in which the teacher starts to re-examine, re-plan, revise or reflect on their original plans, modifying them in some way (SHULMAN 1996 *apud* MIZUKAMI, 2006). Thus, Camila's story may have contributed to Júlia's reflections.

Reflection serves as a starting point for teachers' educational intentions. It is necessary to understand the relationship between content, educational intentions and context (SALAZAR, 2005).

This situation is similar to what Pimenta (1999) says when he states that teacher training in a reflective way represents self-education, because teachers rework their initial knowledge in confrontation with their practical experiences in school contexts. And in this confrontation, in a collective process of exchanging experiences and practices, teachers build up their knowledge, reflecting *on* and in practice:

> He also understands that training is, in fact, self-education, since teachers rework their initial knowledge in confrontation with their practical, everyday experiences in school contexts. It is in this confrontation and in a collective process of exchanging experiences and practices that teachers build up their knowledge as *practicum*, i.e. those who constantly reflect *on* and in practice (PIMENTA, 1999, p.11).

At the next meeting, we worked on reading the text "Talking with your hands". When we asked the participants what they thought should be different for a deaf student, Júlia replied: *"The teaching methodology"*. Below is an excerpt from the text:

> A person who has grown up without listening learns by observation. They need to see, assemble and understand concepts in a concrete way, says Roseli. This is how

> she was able to teach with visual lessons and tangible examples. She used rubber material, squares, cubes, games, dice and drawings. She taught addition with objects that were grouped together. For multiplication, she divided the pupils themselves into squares drawn on the floor: three groups of four equaled 12, five groups of five children resulted in 25. Fractions were understood with circles drawn on the table in the shape of a pizza: with two pieces of the total of eight, a quarter is made. Even probability became easier with a paper doll and various outfits to match (RODRIGUES, 2009, p.2).

Our aim with this text was to show the participants some possibilities for teaching mathematics to the deaf. Julia, despite answering objectively, was able to grasp the main subject of the text.

In addition, she realized that, in order to teach a class with deaf students, it was essential to change the teaching methodology, which, for us, based on our studies, may represent the mobilization of disciplinary knowledge (TARDIF, 2002).

Reflecting on the activity in which she used fraction cubes, she said: *"Difficulties with learning fractions are present at all levels of education. In this activity, my greatest difficulty was in knowing or remembering the mathematical concepts"*.

Julia's perceptions bring us back to the importance of mathematical knowledge as described in the literature review. According to Borges (2013), when the Libras interpreter performs the role of math teacher (without having such training), conceptual errors can occur, making it difficult for deaf students to learn mathematical concepts correctly. Content knowledge is directly related to the subject being taught. While full mastery of the specific content expands the possibilities for teaching intervention, its deficiency limits the paths that teachers can follow when teaching their students (CALDERHEAD, 1988; GROSSMAN *et al.*, 1989; SCHEMPP *et al.,* 1998; SCHINCARIOL, 2002; SIEDENTOP, 2002 *apud* GIRALA; MARCON; NASCIMENTO, 2010). However, just mastering the content is not enough to teach it to any student, including deaf students. Júlia realized that the teacher: *"Has a very important role, because he is the protagonist of the process, not that the student is a supporting actor"* (Field notes, 05/12/15). According to her:

> *Teachers should look for new and diversified methodologies, because students are different and the important thing is learning. Teachers need more incentive to study, because by improving they can improve their work, increasing the level of learning of their students in the classroom* (Field notes, 05/12/15).

Like Paixao (2010), we observed in our study that the knowledge of theory and practice[36] is fundamental to the formation of a teacher capable of dealing with the different situations encountered in professional life.

As we reported earlier in Chapter 4, various aspects of school inclusion were discussed during the course. However, it was only when Júlia watched the video of Camila's interview that she seemed to associate the readings and experiences with the practical life of a deaf student.

In the last activity of the course, Júlia presented a practice using Tangram. Before starting the activity, she explained: *"Students need to have previous knowledge of squares in order to do this activity. If the student doesn't know the square, it's necessary to carry out an activity with this objective"*, because *"[...] it's from something that we start working"*. And she said: *"I planned this activity based on the Christmas Star construction activity carried out in this course"* (Júlia, field notes, 12/12/2015).

According to Pena (2001), the processes by which teachers construct and articulate the knowledge involved in teaching practice highlight the influence of the teacher's personal and professional biography in shaping their pedagogical practice. In this sense, among the types of knowledge involved in teaching practice, pedagogical content knowledge stands out, which refers to the knowledge that teachers have about the content they teach, as well as the way in which they expose this knowledge to a teaching model that produces understanding in students (PENA, 2011).

Júlia's comments indicate that the mathematical content of her activity involves an organization and how this helped Júlia so that the other course participants could understand the ideas, objectives

---

[36] Knowledge from initial training and knowledge from teaching practice.

(general and specific) and structures of the content. We think it's coherent to say that the organization of the content can also constitute pedagogical content knowledge.

Afterwards, Júlia proposed some challenges (e.g. building a cat with the seven handles of the Tangram, building a house shown on the *slide* using the seven handles, etc.). At the end, Júlia said:

> *The interpreter, as she* [Viviane] *said, is a help in the classroom, but he's not the one who has to take on all the responsibility. He's going to help me, be my right and left arm. The responsibility lies with the class teacher. We have to be aware that the support teacher has to support and do so* (Field notes, 12/12/2015).

We notice that the teacher recalls what she said when she watched Camila's video: *"The interpreter is a bridge between you and the student"*. She again reflects on teaching practice. It is important that teachers not only express their thoughts through these reflective activities, but use their reflections to build and revise ideas and knowledge. In this way, teachers can find these activities a useful tool for improving their connections and teaching, so that they can continue to use them in practice (CASTRO, 2009). It is in these reflective processes, in which beliefs, implicit theories and other ways of interacting with contextual conditions define the actions taken to promote teaching in the classroom (SHULMAN, 1987 *apud* DIAZ, 2014).

In her evaluation of the course, Júlia wrote that she felt at ease when she took part, as it enabled her to broaden her vision of her work. She considered visibility, cost and creativity to be important aspects. She reported that she would take the activities she did on the course into the classroom, because after each activity, through reflection, she could see how interested people were in taking part. She said that she would still feel apprehensive about receiving a deaf student in a regular class, but that she would study more to be able to work with the whole class in an inclusive way, with the aim of promoting learning for everyone.

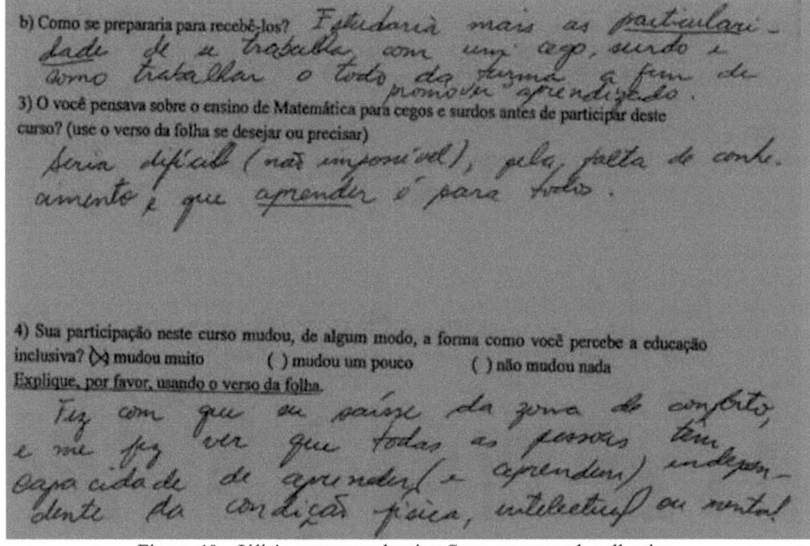

b) Como se prepararia para recebê-los? Estudaria mais as particularidade de se trabalhar com um cego, surdo e como trabalhar o todo da turma a fim de promover aprendizado.

3) O você pensava sobre o ensino de Matemática para cegos e surdos antes de participar deste curso? (use o verso da folha se desejar ou precisar)

Seria difícil (não impossível), pela falta de conhecimento e que aprender é para todos.

4) Sua participação neste curso mudou, de algum modo, a forma como você percebe a educação inclusiva? (X) mudou muito ( ) mudou um pouco ( ) não mudou nada

Explique, por favor, usando o verso da folha.

Fez com que eu saísse da zona de conforto, e me fez ver que todas as pessoas tem capacidade de aprender (e aprendem) independente da condição física, intelectual ou mental.

Figure 40 - Júlia's course evaluation Source: research collection.

Júlia mentioned that her perception of inclusive education had changed, because the activities made her step out of her comfort zone and see that all people have the capacity to learn (and do learn) regardless of their physical, intellectual or mental condition.

In relation to the activities aimed at teaching mathematics to the deaf, Júlia didn't find them very difficult, but she noted that they were challenging. She said that the silence helped her concentrate on the activities, but not being able to speak made it more difficult for her to interact with the environment. As a suggestion, she wrote that the course should have been longer so that she could better develop the proposed activities.

In summary, this experience has shown us that it is relevant and productive to involve teachers in continuing education proposals which allow them to experience situations, exchange experiences, study and, above all, create teaching proposals in line with Inclusive Mathematics Education, which contributes to their professional development. We sought to identify the components of pedagogical content knowledge mobilized by Júlia when carrying out activities related to teaching Mathematics to deaf students in an Inclusive School.

It's interesting to note that the literature review was fundamental to this process. The studies we carried out provided the basis for significant moments in Júlia's career and gave us the opportunity to experience situations that we only knew about from our studies.

Chapter 6

My career as a trainer in training

> *Even based on recent literature that has discussed and explained teacher learning processes, we are faced with a peculiar situation in training agencies. The teacher trainers, whose training and activities are typical of the concept of training processes guided by the paradigm of technical rationality, are the same ones who will provide training processes from a new perspective. [...]* (MIZUKAMI, 2006, p.7).

Like Mizukami (2006), we understand trainers to be all the professionals involved in the formative processes of learning to teach future teachers or those who already work as teachers (teachers of subjects related to teaching practice and supervised internships, pedagogical subjects in general and specific subjects from different areas of knowledge, and professionals from the schools that welcome future teachers). Teacher training is a continuous process of self-education involving individual, collective and organizational dimensions. This process takes place in different contexts and at different times and in different learning communities made up of other trainers (MIZUKAMI, 2006).

Throughout this study, we experienced being a teacher learning to research and, at the same time, learning to be a teacher trainer. It was an intense, difficult process that required a lot of effort. However, it was a rich experience.

Here we will analyze the process experienced by the author of this research[39] , using excerpts from the aforementioned memoir as data, as well as the data produced in this research, especially the researcher's field diary. We have chosen to quote the excerpts from the memoir, the field diary and other records, rather than simply incorporating them into the paragraphs. This choice seemed more appropriate for the analysis.

I have a degree in Mathematics and a Specialization in Mathematics. As I mentioned in the introduction, I've been teaching in public schools for over ten years. I have also taken part in courses on inclusive education. However, these courses were more focused on identifying people with special needs (Down's syndrome, attention deficit hyperactivity disorder - ADHD - etc.). At the time of this research, I was teaching a deaf student called Camila[40] . As in the situation described in the introduction to this research, I was trying, with the support of the Libras interpreter, to provide an opportunity for Camila to learn mathematics (Appendix 1, p.152).

Before I started my Master's degree, I understood that the role of the math teacher was just to be in the same environment as the student. At the time, inclusion seemed very limited to me. Although I was looking for knowledge to get closer to deaf students, I wasn't aware of the complexity of this work. When I taught deaf students, in some cases they were always accompanied by an interpreter. In this way, I thought that, for the teacher, interaction with deaf students was an optional matter that depended on the teacher. I also believed that teaching mathematics to deaf people didn't need to contain the "hardest" content, nor should it be in-depth, because they might not be able to understand it.

As I read throughout my Master's degree, especially the literature review for this research, I began to reflect on what I meant by school inclusion. I understood that school inclusion implies a change in educational perspective. Everyone has the right to education, regardless of whether or not they have a disability. Pedagogical practices must be organized based on the needs of all students. The aim in this context is to involve all students in mainstream education and bring about an improvement in teaching. The interpreter's role is to help deaf students communicate and interpret content. The construction of concepts is the teacher's role.

Therefore, in my opinion, the interaction of the math teacher with the interpreter and the deaf

---

[39] This analysis refers specifically to the experience lived by the author of this research, Viviane, throughout her Master's degree and, in particular, during the extension course. We will therefore use the first person singular in this chapter at all times when reporting moments from this experience.

[40] Fictitious name.

person is not optional, but obligatory. Deafness cannot be an obstacle to deaf education. With appropriate methodologies, deaf students are able to understand any subject they wish. Today, I think like Mendes (2016):

> [...] this tendency to focus on difficulties can be associated with normalization, which, at least implicitly, treats deafness as a complicating factor that prevents or hinders deaf learners' access to mathematical practices determined by 'normals'. [...] Obstacles relating to the educational system, such as the lack of teacher preparation and the absence of information on pedagogical practices in tune with deaf culture, as well as the challenges associated with conducting math classes bilingually, were also cited (p.38).

I also realize that, like many teachers, for a long time I associated inclusion with integration (SASSAKI, 2002).

> By distorting/reducing an idea, we tend to divert ourselves from the challenges of effectively changing our aims and practices. The indifferentiation between the process of integration and that of school inclusion is proof of this trend in education and is reinforcing the validity of the traditional paradigm of educational services. Many, however, continue to uphold it by advocating inclusion! (MANTOAN, 2003, p.14)

Although I did my best to ensure that the deaf students learned mathematics, I didn't think they would get very far in their studies, because the ones I met were very behind in their content and their writing in Portuguese was very poor. I also noticed that, in other subjects, the teachers didn't do any kind of intervention with these students. They were assessed on written work copied from their classmates. I also met a deaf Libras instructor. She had a degree in Higher Normal Education but, according to her classmates, she had never done a written activity or assignment. Her monograph was done by an interpreter. Because she was deaf, she didn't even have to present it.

According to Borges (2013), deaf students find it very difficult to understand the wording of activities in Portuguese and when it comes to reading and interpreting:

> In the case of the deaf, the issue becomes more complicated. We often hear included deaf students showing a lack of interest in reading and interpreting activities, as we have seen in our observations. It has to be considered that the attitudes of teachers who have these students present do not always take linguistic differences into account when selecting activities that increase the difficulties of understanding written texts (BORGES, 2013, p.106).

However, at EBRAPEM in 2015, I met a deaf master's student and my conception was broadened. He spoke and wrote Portuguese correctly. So, right away, I wanted to know what his story was in relation to his studies. He is not totally deaf and has had better opportunities to study. His mother was always very concerned about his education and she was well-off. So that he wouldn't be at a disadvantage compared to his other classmates, she had hired a private Portuguese teacher to help him. This master's student's project was about how deaf math teachers signal certain mathematical concepts. This made it clear to me that there were deaf math teachers.

> Successful deaf students in school mathematics have received very little attention in the literature in contrast to the great emphasis given to the difficulties they experience. It is much more common to find works in which their academic performance is described in terms of failure rather than success (MENDES, 2016, p.37).

And not just math teachers, because a colleague of mine was teaching at a specialized school for deaf students, where almost all the teachers are deaf, including a deaf Portuguese teacher with a PhD in Education. This greatly changed my view of inclusion. Until then, I had studied a lot about teaching the deaf, but I didn't know of any deaf person who had gone beyond graduation knowing how to read and write Portuguese without difficulty. However, the reality was much more complex and multifaceted. I learned that

> [...] the education of Deaf people must take into account that, before being Deaf, they are individuals, and as individuals they must grow and develop, which includes appropriating a language. There is no reason to believe that the language acquisition process for Deaf people is any different from that of their hearing peers (ASSIS,

2013, p.25).

Fortunately, after a lot of study and a few experiences, I realized that for a deaf person to be able to continue their studies, the opportunities they are offered make a difference. Inclusion, in general, is no different. Teachers often don't invest in new methodologies because they believe that students with special needs aren't capable of learning. Today, according to the concept of school inclusion proposed by Mantoan (2005): "[...] inclusion is being with, is interacting with the other" (p.24), I realize that inclusion cannot be limited. We can't just choose the most basic or easiest content to pass on to deaf students. It's very common to prioritize the teaching of geometry because it involves visuals. However, some teachers forget that these students may have the prospect of taking the entrance exam. That's why we think it's plausible for teachers to plan their teaching practices to include all the content corresponding to these students' schooling, and not just geometry. This limitation can deprive students of the opportunities to which they are entitled and leave them on the margins of society.

According to our studies, Pimenta (1999) and Tardif (2002), teaching practice needs to take into account the social context in which the student is inserted. Thus, pedagogical knowledge can be constructed or reinvented based on the pedagogical needs imposed by society. From this perspective, then, the knowledge that the teacher possesses in relation to the social context of his student can help him in his teaching practice and, consequently, in the development of teaching. In this way, the situation described above can prove the mobilization of pedagogical knowledge to build inclusive practices.

School inclusion, in addition to representing an educational turnaround, extinguishes any kind of categorization and stimulates a change in the attitude of the parties that make up this new type of education.

> The educational turnaround that we are proposing through inclusion is based on living together, recognizing and questioning difference and learning as a relational, participatory and meaningful experience for the student. From an institutional point of view, inclusion requires the abolition of categorizations and binary oppositions between students: equal/different; special/normal; healthy/ill; poor/rich; white/black, with mild/severe degrees of impairment, etc. On a personal level, inclusion causes articulation, flexibility and interdependence between the parts that conflict in our thoughts, actions and feelings when we encounter the Other (MANTOAN, 2013, p.l).

In other words, school inclusion proposes a teaching practice that is adapted to all students, based on dialog, respect and participatory and meaningful learning (making sense to the student) through diversified planning and methodologies, developed collaboratively between all its members, including the family and the students. In this sense, the inclusive school represents the student's insertion into the world based on social, cultural and scientific knowledge, regardless of any standards set by society.

However, most of the teachers I worked with were never involved in inclusion. Whenever I taught deaf students, I thought I should do my best to teach them, because if they were my family, I would want them to be treated that way. For me, teachers need to plan their practice with the particularities of their students in mind. They also need to be able to provide ways for all students to understand and apply knowledge.

> Inclusion does not require the use of specific school teaching practices for this or that disability and/or learning difficulty. Students learn within their limits and if the teaching is really good quality, the teacher will take these limits into account and exploit the possibilities of each student appropriately. It is not a question of passively accepting school performance, but of acting realistically and coherently and admitting that schools exist to train the new generations, and not just some of their future members, the most capable and privileged (MANTOAN, 2002, p.4).

Today, when it comes to school inclusion, I believe that the teacher can make a difference. The more the teacher remains calm, persevering and willing to help all students, the greater the chance of success in their teaching practices. The more the teacher gets involved in building a learning environment that is favorable to all students, the greater the chance of generating good results for both the students and themselves.

> [...] day-to-day teaching work is fundamentally a set of personalized interactions with students in order to get them involved in their own learning process and meet their different needs. This is why this work constantly requires a deep investment, from both an affective and cognitive point of view, in human relationships with students. These relationships can hardly be superficial. They require teachers to be personally involved in the interactions, because they are dealing with people who can deviate and annul the work process in different ways, and from whom they must obtain assent or consent, and even participation. In this sense, the teacher's personality is an essential component of their work. [...] a teacher cannot only 'do his job', he must also commit and invest in that job what he is as a person (TARDIF, 2002, p.141).

Like Tardif (2002), I believe that "it is clear that some teachers may well evade this demand and live their work in an indifferent, detached way" (p.141). However, in doing so, they renounce the very meaning of the profession. For this author, the teacher, with all that he is (qualities, defects, sensitivities, emotions and affectivities), becomes, in a way, a technological component of the interaction professions. And this emotional technology is represented by physical postures and ways of being with students. This is because "the personality of teachers permeates pedagogical practice: there is no objective or general way of teaching, every teacher transposes what they are as a person into their practice" (TARDIF, 2002, p.144).

During my Master's degree, I had the opportunity to observe different positions on inclusion. A colleague, who taught in a specialized school for the deaf, believed that this was a more appropriate situation for deaf students, because they respected each other, had the same language and were part of the same culture. I didn't agree because, for me, deaf people couldn't be limited to their own culture. I saw the inclusion of deaf students as an opportunity for socialization that would benefit everyone, deaf and hearing. Like Neves (2011), I believed

> that the pedagogical strategy of bilingualism must be the basic center of mathematics teaching for this clientele, which must be understood and thought of through the lens of a culture, of its own identity to be taken as relevant when thinking about specific and appropriate teaching conditions. Mastery of mathematical content or even articulation with Libras does not guarantee efficiency in teaching mathematics to these individuals (p.112).

However, we both learned a lot. I had the chance to come into contact with unfamiliar realities and to reflect on my convictions. And my colleague also realized some of the limitations that existed in a special school, because some of its deaf students didn't have much idea of what was going on in the world, they didn't have access to the news and some didn't even know money. However, in that same school, there were some deaf people who preferred to relate only to deaf students, and their parents thought that was better.

In this way, I realized that the inclusion of deaf people can have different perspectives. In our historical retrospective, we saw that deaf people were deprived of education for a long time. The special school improved the educational condition of the deaf. However, I continue to believe that:

> We can't imagine education for all when we fall into the temptation of grouping students by grades, by levels of school performance and determining objectives for each level. And even more so, when we send those who don't fit into any of these groups to special classes and schools, arguing that education for all would not be distorted in such cases! (MANTOAN, 2015, p.40)

So the important thing is to respect people and their opinions on a subject, even if we don't share the same ideas. From this situation, I realized that I should deal in the same way not only with my students, but also with all my co-workers.

Consequently, the way I perceive teacher training (initial and continuing) to work in an inclusive school has changed a lot. Before the Master's and even during the fieldwork, I believed that if I told people about my experience with deaf students, teachers or future teachers would be able to think and plan their lessons in an inclusive way. I didn't believe in diversifying methodologies. For me, the blackboard and chalk were more than enough. It was all a question of teacher awareness.

However, after my Master's degree, I realized that training (in person and continuing) could be offered in a more reflective way, with the teacher occupying a more autonomous space[41] . I realized that teachers need to be encouraged and valued. Training cannot be an imposition of knowledge. It needs to take into account the experiences and knowledge that teachers already have; provide dialogue, offer and stimulate the updating of knowledge and propose activities that prioritize the experience and construction of inclusive practices aimed at teaching mathematics to deaf students in mainstream schools.

> Both in its theoretical basis and in its practical consequences, professional knowledge is evolving and progressive and therefore requires continuous and ongoing training. Professionals must therefore self-train and retrain through different means after their initial university studies. From this point of view, professional training occupies, in principle, a good part of the career and scientific and technical knowledge has the property of being revisable, criticizable and subject to improvement (TARDIF, 2002, p.249).

At the time of the extension course, I was teaching in two public secondary and elementary schools. I had never taught teachers or future teachers before. This made me a little apprehensive at first.

Despite having professional experience as a math teacher and even teaching classes with deaf students, I felt insecure about this new situation. The prospect of "teaching" teachers reminded me of conversations with other math teachers about my practice when I taught deaf students. They usually told me that it was impossible and that they weren't prepared for it.

Today, I see that this reaction is common among teachers. This is evident in the literature. Spenassato and Giareta (2009) investigated how math teachers developed their lessons when teaching deaf students in mainstream classes. The reports of the study participants show that none of them changed their teaching practices in deaf and hearing classes, because they felt unprepared in terms of their choice of methodology.

So, although I acted differently in my own teaching practice when I started receiving deaf students, I knew it wouldn't be easy to work with teachers in the extension course. Despite my fears, like Santos (2012), I believe that

> [...] the inclusion of deaf students in math classrooms requires much more than learning on their part. Equally important is that we, as mathematics educators, have to learn, and value, different ways of doing mathematics and that we have to fight for a mathematics curriculum that reflects the diversity of our students, a curriculum that recognizes that there are various ways of doing mathematics and various ways of appropriating it (p.121).

So, together with my advisor and fellow master's student, I tried to think of the extension course as an opportunity to involve teachers and future math teachers (as well as the other participants) in situations that would allow them to rethink inclusion in math classes.

> After studying for my master's degree, I realized that training (in-person and continuing) should be offered in a more reflective way, where the teacher is more autonomous. I realized that teachers need to be encouraged and valued. Training cannot be an imposition of knowledge. It needs to take into account the experiences and knowledge that teachers already have. Create study groups and encourage dialog. Propose activities that prioritize the experience and construction of inclusive practices aimed at teaching mathematics to deaf students in mainstream schools. In these activities, teachers should be able to experience the limitations of not being able to hear, so that they can begin to be aware of the learning conditions of deaf students. It should broaden teachers' didactic knowledge, especially with regard to Inclusive Schools. Stimulate a change in the teacher's profile so that they add a new culture to their practice, based on dialog, reflection and collectivism (Memorial, 06/11/16).

This proposal is close to Pimenta's (1999) ideas about teacher training in the reflexive trend.

---

[41] Stop reproducing the same activity several times and create their own inclusive teaching practices based on readings or related activities aimed at the particularities of their students.

This tendency is characterized as a policy of valuing the personal-professional development of teachers and school institutions, suggesting working conditions that foster continuous training in the profession, creating networks of self-training.

An example of this appears in the reading of the text "Talking with your hands" (meeting, 28/11/15).

> Rather than showing the participants ready-made practices, I thought it would be interesting to point out that teachers can build their own practices in this context. I thought this article could motivate the participants to create their own inclusive practices (Viviane, research collection).

According to Pimenta (1999), teaching is not a bureaucratic activity for which knowledge and technical-mechanical skills are acquired. The teaching job consists of teaching as a contribution to the process of humanizing students according to the context in which they are inserted. For the author, being a teacher involves developing knowledge and skills, attitudes and values that enable them to permanently build their teaching know-how based on the needs and challenges that teaching, as a social practice, poses to them on a daily basis. Teachers are expected to mobilize the knowledge of educational theory and didactics needed to understand teaching as a social reality, and to develop the ability to investigate their own activity in order to build and transform their teaching know-how, in a continuous process of building their identities as teachers (PIMENTA, 1999).

In this way, initial or continuing training represents self-training, as teachers rework their initial knowledge in confrontation with their practical experiences in school contexts. In this confrontation, in a collective process of exchanging experiences and practices, teachers build up their knowledge, reflecting *on* and in practice. Reflective teacher training values the personal-professional development of teachers and school institutions, because it works with knowledge in the dynamics of a multi-media society, globalization, multiculturalism, transformations in productive markets, and in the training of pupils, children and young people (PIMENTA, 1999).

It is interesting to note that my considerations about training in a reflective way are also close to the ideas of Mantoan (2003) in relation to initial and continuing training aimed at school inclusion. According to the author, teaching from an inclusive perspective involves *re-signifying* the role of the teacher, the school, education and pedagogical practices in order to overcome the exclusionary context of our education at all levels. Thus, the author states that preparing teachers to teach in Inclusive Schools involves a different *design* from existing professionalization proposals. It involves questioning one's own practice, making comparisons, analyzing the circumstances and facts that cause disturbances and/or are responsible for school success, where teachers define their "pedagogical theories".

As a trainer in the making, I've been learning the importance of questioning my own practice, carefully analyzing the context in which I work and getting to know the people involved in that context. In a way, I have my own "pedagogical theories".

In this sense, I believe I have mobilized knowledge related to various domains, but one stands out: pedagogical content knowledge. In this context, this knowledge is the only type of knowledge in which the trainer is really the protagonist and includes knowledge about the promotion of teaching learning processes and about professional practice as the axis of training processes (MIZUKAMI, 2005). Through this, the trainer will be able to assess the extent to which they are achieving their objectives and choose which paths to follow if their training proposal is not progressing as expected.

In my reflections, I notice the effort to build a learning environment in which mathematical concepts were worked on in such a way that the course participants could both remember/learn the topics in question and realize that it was possible to do this with all students, including deaf ones, and in an interesting way. An example of this is the star activity:

> I started by looking for a Christmas ornament that was easy to make, low cost and that the participants could extract mathematical content to work on with deaf and hearing students. As in this proposal the participants were only supposed to reproduce the ornament by observing its construction stages, I soon did some searching on the internet for videos of Christmas ornaments. I prioritized not only the number of mathematical concepts, but also the beauty of the ornament so that

> the participants could use it in their homes. So I chose the Christmas star made from laminated paper, which was very eye-catching and made it easy to visualize, which, according to the literature review, can facilitate the teaching of mathematics to the deaf, as well as having various mathematical concepts to explore, since the aim was not simply to make an ornament. The aim of this activity was to teach mathematical concepts in a different and useful way. [...] In this activity, I tried to communicate with the participants in the same way that I communicate with my deaf students in the classroom, writing down one word at a time to ask a question and, to reinforce it, underlining it a few times. As the participants' language was Portuguese, I wrote some of the questions with all their elements (article, preposition...) on the board, underlining verbs, nouns and adjectives, following the characteristics of Libras, which has no meaning for articles, conjunctions and prepositions. When I taught deaf and hearing students in the same classroom, I did it this way (Viviane, research collection).

I also tried to articulate knowledge (among them: mathematical knowledge, knowledge of professional practice and knowledge evidenced by the literature review) that would favor teacher learning.

When developing the activity, I was thinking as a teacher - trying to present an interesting proposal that would allow all the students to learn mathematical concepts - and as a trainer - providing a rich experience for all the participants, despite their heterogeneity, in which they would improve their mathematical knowledge in a playful activity and, above all, realize that it could be done with all the students. These concerns are similar to those put forward by Pimentel (2012) when he states that adapting the curriculum to ensure that it takes account of the diversity in the classroom involves reflecting on the proposed curriculum, questioning the existing content and previously defined objectives, taking the reality of the whole class as a yardstick. This does not mean that each teacher will create a new curriculum based on the actual development observed in a class. Adapting the curriculum does not mean proposing a different curriculum, but adapting that content, that time allotted for learning to the reality of their group of students. Therefore, like Aranha (2004), I understand that:

> With regard to the pedagogical proposal, it is important to point out the importance of making the curriculum more flexible in order to make the inclusion process viable. If they are to be facilitators and not hindrances, curricular adjustments need to be thought out from the group context in which a particular student is placed. Adaptations refer to a context, and not to the child, to the particular meeting point that occurs in the classroom, where the child, their history, the teacher, their experience, the school institution, the curriculum plan, the regulations, the parents' expectations, among others, converge. It is therefore not possible to think of general adjustments for children in general. Curricular flexibilization must be considered on the basis of each particular situation, and not as universal proposals, valid for any school context. The adjustments made by a particular teacher for a specific group of pupils are only valid for that group and that moment (p.20).

As the extension course, from the second meeting onwards, was developed on the basis of the participants' reflections, evaluations and suggestions, each week I expanded my knowledge in an effort to plan and develop enriching meetings for the participants.

In terms of training strategies, I learned a lot. I would particularly highlight the use of teaching cases. I think that the interview with Camila was a variation on this strategy, as it gave the participants a glimpse into the experience of a deaf student.

> I decided to ask the interpreter if my student Camila [at the time] would like to collaborate with my research. When I met her at the beginning of the year, she didn't like math, had a sad face and reported many difficulties in learning. And only a few changes in the planning of math classes changed Camila's outlook on her studies. So I decided to interview Camila, because there was no one better than her to explain to the participants the difference a teacher can make to a deaf student's school life (Viviane, field notes).

Teaching cases are generally used with a view to: developing reflective processes at different times and levels; analyzing manifest conceptions and understanding specific learning in concrete teaching and learning situations; objectifying and discussing beliefs, as well as explaining practices based on the situation studied; and constructing situations that enable processes of reflection on action (SCHÓN, 1983 *apud* MIZUKAMI, 2005).

In this sense, Aranha (2004) emphasizes that the development of an inclusive school involves reflection, the organization of actions and the participation of the entire school community in a collective process of construction. In this way, as everyone is involved in reflecting on the school, on the community from which its pupils come, on the needs of that community, on the objectives to be achieved through educational action, the school will come to be felt as it really is: of everyone and for everyone (ARANHA, 2004).

Camila's interview allowed me to explore my history as a math teacher in order to achieve some objectives as a trainer. With Camila's interview, I sought to promote discussions and trigger reflective processes on the subject, essential characteristics for a teaching case. At the same time, I was reflecting on these processes myself:

> Julia said twice that inclusion was a matter of respecting differences and that she didn't believe in inclusive schools. And when talking about the teacher's relationship with the deaf student, she reported that it went smoothly. The situation experienced by Camila throughout her student life does not seem to us to be any different from the situation of the teacher reported by Julia. In our opinion, the situation has only changed narrator. However, when we watched the interview, Julia didn't signal the same way. According to Shulman (1996), stories that start from experience are transformed into cases through narration. And the plot of each case involves a context, an intervention that is not complete, making it possible to create situations in which the teacher begins to re-examine, re-plan, revise or reflect on their original plans, modifying them in some way (SHULMAN 1996 *apud* MIZUKAMI, 2006). Thus, Camila's story prompted Julia's reflection on practice in the inclusive school (Field diary).

I learned a lot at each meeting, putting into practice everything I had studied and planned, in partnership with my advisor and my fellow master's student. In practice, when developing the tasks with the course participants, new questions always arose that made me reflect on teaching learning (mine and theirs) and on the teaching and learning of mathematics from an inclusive perspective.

I clearly understood what Cochran-Smith (2003 *apud* MIZUKAMI, 2005) meant when he said that teachers are students of their field throughout their lives, and that this takes place in both individual and collective dimensions.

> Two basic focuses should be considered in a community of teachers: the improvement of professional practice and the belief that teachers are lifelong learners of their fields. As such, they must grow in knowledge, breadth and understanding, as well as keeping up with the process of knowledge production in their specific fields (COCHRAN-SMITH, 2003 *apud* MIZUKAMI, 2005, p.13).

Another aspect worth highlighting is the role of an investigative attitude on the part of teachers and trainers. I see this as essential in the exercise of the teaching profession and in the work of the teacher trainer. This investigative attitude involves an intellectual perspective, a way of questioning, making sense of and relating daily work to the work of others and to wider social, historical, cultural and political contexts (COCHRAN-SMITH, 2003 *apud* MIZUKAMI, 2005).

Although I realize the limitations of the analysis carried out - many of them characteristic of researchers in training carrying out their first investigation - I believe it was a worthwhile effort. As well as being the researcher in this work, I acted as a trainer in the extension course without ever ceasing to be a math teacher. Experiencing these three identities simultaneously was sometimes a challenge. However, it was a very rich learning experience for me as a professional.

FINAL CONSIDERATIONS

> *Research on the reflective teacher, by placing the links between training and profession as constituents of the specific knowledge of teaching, as well as the material conditions in which it is carried out, values the work of the teacher as the subject of the transformations that are necessary in the school and in society* (PIMENTA, 1999, p.30).

We began this research with the desire to improve math teaching practices for deaf students in order to favor their learning. To this end, we set out to invest in teacher training, seeking to sensitize them to inclusion and make them reflect on this scenario in such a way as to perceive themselves as capable of acting in this direction, believing in the potential that all students, including deaf students, bring with them.

Within these broad objectives, we worked on a more modest one: mobilizing teaching knowledge, especially knowledge related to the inclusion of deaf students in mathematics classes through participation in an extension course.

Like Souza (2013) we believe that:

> At the moment, after the government initiatives guaranteeing inclusion, in this case that of deaf students, it is up to the teacher to create and develop the right conditions to provide care that respects their specificities. However, such actions will only be possible through processes that allow the teacher to plan and create possible action strategies through what is understood as curricular adaptations (p.82).

Based on our readings and reflections on school inclusion, teaching knowledge and the teaching and learning processes of Mathematics for deaf students and the difficulties they encounter, we outlined our object of study: the teaching knowledge of teachers who teach Mathematics, more specifically the teaching knowledge related to the inclusion of deaf students mobilized by a study group.

With this in mind, this work was carried out with the following research questions in mind:

How does participating in an extension course aimed at including deaf people in mathematics classes contribute to the mobilization of teaching knowledge on the part of a mathematics teacher? How does this experience contribute to the mobilization of teaching knowledge on the part of the teacher trainer?

The results, in general, show that all the participants got involved - to a greater or lesser extent - with the activities carried out during the course and that they all became more aware of inclusion, broadening their perception of what a deaf student is capable of learning, as well as the possibilities of teaching mathematics to classes involving deaf students.

The analysis was based on two perspectives on the process: that of a teacher who took part in the course and that of the trainer/researcher. There are indications that both, in different ways, mobilized teaching knowledge for inclusion, especially pedagogical content knowledge.

The first, despite being an experienced math teacher, broadened her knowledge by experiencing and constructing mathematical tasks that could be developed with all students, including deaf students. She also reframed her perception of the roles of the interpreter and the math teacher in inclusive classes.

The second, when reflecting on the process of learning to research and to be a trainer, without losing sight of her professional experience as a math teacher, noticed an intense mobilization of knowledge resulting not only from studying and preparing for the extension course meetings, but also from contact with the group. This experience profoundly broadened her perception of teaching from an inclusive perspective and of the importance of self-training for teachers and trainers. Another contribution was the trainer's perception of the importance of the reflective teacher, constituting their knowledge and reflecting on and *in* practice.

We believe that school inclusion involves teachers building up specific teaching knowledge. Like Silva (2014), we realize that reflection on educational practice is particularly important in this process:

> Reflection on educational practice is based on the understanding that it is necessary

> to think about the appropriation and construction of the teaching subject's knowledge, making them capable of knowing the elements of their social context every day, in order to intervene in it, expanding communication, freedom and interaction between individuals. Knowledge, therefore, implies a search for and dialectical articulation between theoretical knowledge and practical knowledge in a context that envisages constant transformation, as well as the emancipation of those involved in this process, whether in professional training or in teaching practice itself (p.59).

We believe that the extension course achieved this goal by stimulating reflection both through the experiences and the discussions that followed them. In addition, the reflections and evaluations produced during the course by the participants favored our own learning, both by providing elements for the construction of activities for the course, more in tune with their interests, and by favoring our understanding of how to act as a trainer by pointing out challenging aspects or those that seemed easy, interesting or not, etc. In this sense, promoting reflections on the meetings orally and in writing proved to be an interesting strategy. We noticed that some participants felt more comfortable writing down their opinions rather than telling them to the whole group.

It is important to emphasize that all the studies carried out during the research were useful both for the development of the extension course and for the interpretation and analysis of the data. The visualization mentioned by some authors proved to be a fundamental aspect for the development of the activities.

It is important to emphasize that, although our research question deals with teaching knowledge, we corroborate Morin's ideas (1993 *apud* PIMENTA, 1999) in relation to knowledge. We recognize that knowledge is not reduced to information. Information is only the first stage. Knowing implies a second stage, that of classifying, analyzing and contextualizing information. The third stage is related to intelligence, awareness or wisdom. Intelligence is the art of linking knowledge in a useful and coherent way. Consciousness and wisdom give us the ability to produce new forms of existence and humanization. In short, knowing means being aware of the power of knowledge (MORIN, 1993 *apud* PIMENTA, 1999).

Our intention was not to exhaust all the discussions regarding the teaching of mathematics to deaf students in inclusive schools, but to promote reflections on an experience of training mathematics teachers in this context.

Even with the limitations and all the difficulties involved in the research process, we see this study as a possible stimulus for teachers or future teachers to create their own inclusive practices for teaching mathematics to the deaf.

REFERENCES

ALONSO, D. **The challenges of Inclusive Education: focus on support networks.** Available at: < https://novaescola.org.br/conteudo/554/os-desafios-da-educacao-inclusiva-foco-nas-redes- de-apolo > Accessed on: May 05, 2015.

ARANHA M.S.F. **Educado inclusiva**, v. 3: a escola / coordenado geral SEESP/MEC. Brasilia: Ministry of Education, Special Education Secretariat, 2004. 26 p.

ASSIS, C. **Exploring the idea of the rational number in its fractional representation in Libras**. 2014.176f. Master's dissertation in Mathematics Education - Bandeirante Anhanguera University, Sao Paulo, 2013.

BARRETEA, E.M. e CANAN S.R. **Políticas Públicas de Educagao Inclusiva** : avangos e recuos a partir dos documentos legais. In:9⁰ ANPED SUL. Seminar on Educational Research in the Southern Region. Caxias do Sul. p.1-15.Jul-Aug.2012.

BARROS, A.A.M.**Teaching knowledge in the context of early childhood education: pedagogical practice in focus.**2016.249f. Master's dissertation in Education - Pontifical Catholic University of Campinas, Sao Paulo, 2015.

BESERRA M.P.S. **Fontes e características dos saberes pedagógicos dos docentes da Universidade Federal do Piauí** - Campus Ministro Reis Velloso - Parnaiba - Available at
<http://leg.ufpi.br/subsiteFiles/ppged/arquivos/files/eventos/evento2002/GT.1/GT1_9_2002.p df> Accessed on 19/03/02017.

BEZERRA, C. **Interaction between deaf learners using the discussion forum**: Limits and Potential. 2013.148f. Master's dissertation in Mathematics Education - Bandeirante University of Sao Paulo, Sao Paulo, 2012.

BOLÍVAR A. **Pedagogical content knowledge and subject matter didactics. Pedagogical content knowledge and subject matter didactics**. Revista de curriculum y formación del profesorado, p.2-9. (2005) Available at: <
https://www.ugr.es/~recfpro/rev92ART6.pdf > Accessed on April 14, 2017.

BORGES, F.A. **Inclusive education for the deaf:** an analysis of mathematical knowledge mediated by the Libras interpreter. 2014.206f. PhD thesis in Education for Science and Mathematics - State University of Maringá, Maringá, 2013.

BRAZIL, (2013). Ministry of Education. Secretariat for Literacy and Diversity. **Report on the Linguistic Policy of Bilingual Education - Brazilian Sign Language and Portuguese Language of the Working Group**, designated by Ordinances No. 1,060/2013 and No. 91/2013 of MEC/SECADI. Brasilia, February 2014. Available at: < http://www.bibliotecadigital.unicamp.br/document/?code=56513 >. Accessed on: 10/12/2014.

BRAZIL, Law No. 13.146; 2015. **Statute of Persons with Disabilities, Brazilian Law for the Inclusion of Persons with Disabilities**. Available at <
http://www.planalto.gov.br/ccivil_03/_Ato2015-2018/2015/Lei/L13146.htm > Accessed on December 11, 2016.

BRAZIL, Ministry of Education and Sports/Secretariat for Primary Education. **Parámetros curriculares nacionais: Adaptagóes Curriculares** / Secretaria de Educagao Fundamental. Special Education Secretariat. - Brasilia: MEC / SEF/SEESP. 62 p.1998.

° BRAZIL, Ministry of **Education/National Education Council Resolution № 2,** of July 1, 2015. Available at: < http://PRONA campo. mec.gov.br/images/pdf/res_ cne_cp_02_03072015.pdf> Accessed on: September 04, 2016.

CARVALHO, Rosita Edler. **Educagáo inclusiva:** com os pingos nos "is". Porto Alegre: Mediagao, 2004.

CASTRO J.L.F. **Conocimiento pedagógico del contenido en la formación** *de* **docentes** *de* **matemática** .*Cuadernos de Investigación y Formación en Educación Matemática.* Año 4. Número 5. pp 11- 27.Costa Rica. 2009. Available at: <

http://revistas.ucr.ac.cr/index.php/cifem/article/viewFile/6912/6598 > Accessed on April 22, 2017.
CAWTHON S. **Professional Development for Teachers of Students Who Are Deaf or Hard of Hearing:** Facing the Assessment Challenge **American Annals of the Deaf** Volume 154, Number 1, p. 50-61, Spring 2009 .
CINTRA, V.P. **Trabalho com projetos na Formagáo inicial de Professores de Matemática na Perspectiva da Educagáo Inclusiva.** 2015.137f. PhD thesis in Mathematics Education - Paulista State University, Rio Claro, 2014.
CONCEIQÁO, K.S. **The construction of algebraic expressions by deaf students:** the contributions of Micromundo Mathsticks. 2013. 127 f Master's dissertation in Mathematics Education - Universidade Bandeirante de Sao Paulo, Sao Paulo, 2012.
COUTINHO M.C.B. **A construgáo** de **saberes docentes para a Inclusão das pessoas com deficiência:** um estudo a partir dos professores do curso de Pedagogia do Sertao. 2013.145 f.; Master's dissertation - Federal University of Pernambuco, Pernambuco, 2012.
CUNHA R.E. **Os Saberes Docentes ou Saberes dos professores.** Available at: < http://www.vdl.ufc.br/solar/aula_link/llpt/A_a_H/didatica_I/aula_01/imagens/03/sabere s_doc entes.pdf> Accessed on: April 10, 2016.
CYSNEIROS P.G.**Competencias para ensinar com novas Competences.**Available at :< www2.pucpr.br/reol/index.php/DIALOGO?dd1=617&dd99=pdf> Accessed on March 10, 2017.
DELLANI M.P and MORAES D.N.M. **Inclusão: Caminhos, encontros e descobertas.** Revista de Educagao do Ideau,Vol. 7 - № 15 - January - June 2012. Biannual ISSN: 1809-6220 Available at: < http://www.ideau.com.br/getulio/restrito/upload/revistasartigos/50_1.pdf> Accessed on: September 05, 2016.
DIAZ C.V. and MARDONES H.C. **Conocimiento Pedagógico del Contenido: ¿el paradigma perdido en La formación inicial y continua de profesores en Chile?**Available at: < http://www.scielo.cl/pdf/estped/v40nEspecial/art19.pdf> Accessed on: April 15, 2017.
DONATO, S. P. and ENS, R. T. **Contemporary teaching:** between teaching knowledge and practices. Available at: < http://www.pucpr.br/eventos/educere/educere2008/anais/pdf/192_353.pdf> Accessed on: July 27, 2015.
DUK, Cynthia. **Educating in diversity: teacher training material**. 3. ed. Brasilia: MEC.SEESP, 2006. Available at :< http://portal.mec.gov.br/seesp/arquivos/pdf/educarnadiversidade2006.pdf> Accessed on: September 06, 2016.
FALCÁO, L.A.B. **Educagáo de surdos e a formagáo docente na perspectiva inclusiva** - Saberes e conflitos educacionais brasileiros. Available at :< http://www.porsinal.pt/index.php?ps=artigos&idt=artc&cat=7&idart=137 > Accessed on: July 21, 2015 .
FERNANDES, S. H. A. A.; HEALY, L.. **Ensaios sobre a inclusão na Educagáo Matemática**, n. 10, p.59-76, 2007. ISSN: 1815-0640. Available at: < http://www.fisem.org/www/union/revistas/2007/10/Union_010_010.pdf > Accessed on: September 06, 2016.
FERNANDEZ C.; **Revisiting the knowledge base and pedagogical content knowledge (PCK) of science teachers;** Revista Ensaio, v.17, n. 2, p. 500-528, Belo Horizonte, May-Aug, 2015.
FILHO, G.A.T and MIRANDA G.T. **O professor e a educagão inclusiva**: formagáo, práticas e lugares, EditoraUFBA. Salvador, Bahia, 2012.
FINIQUETO L.C. P; SANTOS M.E.G. and TERRAZZAN E.A. **Os saberes docentes**

ao desenvolvimento profissional dos professores. Available at: < http://alb.com.br/arquivo- morto/edicoes_anteriores/anais15/Sem13/leilafinoqueto.htm > Accessed on: May 20, 2016.
FRIZZARINI, S.T. **Study of semiotic representational registers:** implications for the teaching and learning of algebra for deaf students fluent in sign language. 2013.288 f PhD Thesis in Education for Science and Mathematics - State University of Maringá, Maringá, 2014.
GARCIA, B.M.M. **Educagáo infantil**: saberes e práticas da inclusáo: introdução. [4. ed.] / prepared by Marilda Moraes Garcia Bruno. - Brasília: MEC, Secretaria de Educagáo Especial, 45 f, 2006.
GARRIDO, E. and BRZEZINSKI, I. Reflection and investigation of one's own practice in initial and continuing training: contribution of dissertations and theses in the period 1997-2002. **Revista Diálogo Educativo**, Curitiba, v. 8, n. 23, p. 153-171, jan./abr. 2008. Available at: <ww2.pucpr.br/reol/index.php/dialogo?dd99=pdf&dd1=1834>. Accessed on: April 22, 2017.
GIL, R. S. A. **Mathematics education for the deaf:** a study of the training needs of teachers who teach mathematical concepts in the context of education for the hearing impaired in Belém do Pará. 2008.191f. Master's dissertation in Science and Mathematics Education - Federal University of Pará, Belém, 2007.
JESUS, T.B. **(De)construction of geometric thinking:** a shared experience between teachers and a deaf student. 2015.183f. Master's dissertation in Science and Mathematics Education - Federal Institute of Espirito Santo, Vitoria, 2014.
MACHADO G.C. **The construction of knowledge necessary for school inclusion: training and experience of teachers in Montenegro/RS**. Available at :< http://www.portalanpedsul.com.br/admin/uploads/2012/Formacao_de_Professores/Trab alhoZ0 2_20_01_16-6892-1-PB.pdf> accessed on April 13, 2017.
MANTOAN, M.T. E. **School Inclusion: What is it? Why? How to do it?** Sao Paulo: Moderna, p. 13-20 and 27-34, 2003. Letter from Professor Mantoan to senators. Available at < https://inclusaoja.com.br/2013/11/07/carta-da-profa-mantoan-aos-senadores/ > Accessed on February 28, 2017. MANTOAN, Maria Teresa Eglér. Pedagogical paths to inclusion. Sao Paulo: Memnon, edigoes científicas, 2003. Available at < http://www.educacaoonline.pro.br/index.php?option=com_content&view=article&id=8 3:cami nhos-pedagogicos-da-inclusao&catid=6:educacao%20inclusiva&Itemid=17/ > Accessed February 28, 2017.
MARCON D, GRAQA A.B.S, NASCIMENTO J. V. **Reflexoes sobre o processo de construgao do conhecimento pedagógico do conteúdo de futuros professores**. Available at < http://www.scielo.br/pdf/refuem/v23n2/14.pdf >Accessed on April 14, 2017.
MENDES R.G. **Successful deaf people in mathematics**: relationships between their cultural values and their mathematical identities. 126f. Master's dissertation in Mathematics Education - Bandeirante Anhanguera University, Sao Paulo, 2016.
MINAYO M.C.S.Pesquisa Socila : teoria método e criatividade.30ª ed. Petrópolis . Editora Vozes,2011.p.60-110.
MIZUKAMI M.G.A.**Teaching learning**: teacher trainers. Revista E Curriculum, Sao Paulo, v. 1, n. 1, dec. - jul. 2005-2006. Available at < https://revistas.pucsp.br/index.php/curriculum/article/viewFile/3106/2046 > Accessed on April 15, 2017.
NASCIMENTO, P.R. **Uma proposta de formagao do professor de matemática para educagao de surdos.** 2010.89f. Master's dissertation in Mathematics and Science Teaching - Federal Center for Technological Education Suckow da Fonseca CEFET/RJ, Rio de Janeiro, 2009.

NETO A.R. **Pessoa com deficiência, necessidades especiais e processo-** apontamentos acerca dos mecanismos de igualdade e o projeto do novo código de processo civil, 2012. Available at: < https://por-leitores.jusbrasil.com.br/noticias/3154136/pessoa-com-deficiencia -necessidades- especiais -e-processo -apontamentos- acerca- dos-mecanismos-de igualdade-e-o-projeto-do-novo-cpc > accessed on January 13, 2017.
NEVES, M.J. **Mathematical communication in the classroom**: methodological obstacles in the education of deaf students. 2012,131f. Master's dissertation in Mathematics and Science Education - Federal University of Pará, Belém, 126f. 2011.
NÓVOA, A. **Teacher training and the teaching profession**. In: NÓVOA, A. (Coord.). Os professores e a sua formagao. Lisbon: Dom Quixote, 1995.
NUNES, C.M. F. **Saberes docentes e formacao de professores: um breve panorama da pesquisa brasileira**. Educagao & Sociedade, ano XXII, n^0 74, Abril/2001. Available at: http://www.scielo.br/pdf/es/v22n74/a03v2274.pdf> Accessed on April 23, 2017.
OLIVEIRA, C. L. **A theoretical and conceptual overview of qualitative research:** technical types and characteristics. Revista Travessias. Educagao, Linguagem e Artes,v. 2, n. 3,ed.04,2008.Disponível em:< http://e-revista. unioeste.br/i ndex.php/ travessias/article /view/3122/2459 > Accessed on: October 06, 2016.
OLIVEIRA, J.S. **A comunidade surda:** perfil, barreiras e caminhos promissores no processo **de** ensino e aprendizagem **de** matemática. 2006.78f. Master's dissertation in Mathematics and Science Teaching - Federal Center for Technological Education Suckow da Fonseca CEFET/RJ, Rio de Janeiro, 2005.
PAIXÁO, S.S.M. **Saberes de professores que ensinam Matemática para alunos surdos incluídos num escola de ouvintes.** 2011.200f. Master's dissertation in Mathematics and Science Teaching - Federal University of Pará, Belém, 2010.
PASSOS, A.A; PASSOS M.M. e ARRUDA S.M. A **Educagao Matemática Inclusiva no Brasil: uma análise baseada em artigos publicados em revistas de Educagao Matemática**. Revista Brasileira de Ensino de Ciencia e Tecnologia**.** Ponta Grossa. v.6, n.2, p.1-22.May-Aug.2013.
PESCE, M.K., ANDRÉ M.E.D. e HOBOLD M.S. **Formação do professor pesquisador:** procedimentos didáticos. Working Group: Teacher Training and Teaching Professionalization. Available at: < http://educere.bruc.com.br/arquivo/pdf2013/7520_5224.pdf > Accessed on April 15, 2017.
PIMENTA, S. G. **Teacher training:** identity and teaching knowledge. In: (Org). Pedagogical knowledge and teaching activity. Sao Paulo: Cortez Editora, p. 15 a 34,1999. Available at: < https://edisciplinas.usp.br/pluginfile.php/1978920/mod_resource/ content/1/Texto-%20Pimenta-%201999-FP-%20ID%20%20e%20SD.pdf > Accessed on March 23, 2017.
PIMENTEL S.C. **Teacher training for inclusion**. Necessary knowledge and training paths. In: O professor e a educação inclusiva formagao, práticas e lugares Available at,<http://www.google.com.br/url?sa=t&rct=j&q=&esrc=s&source=web&cd=1&cad=rj a&u act=8&ved=0ahUKEwixsdygwLvTAhVDCpAKHcn3BiMQFggkMAA&url=https%3A %2F %2Frepositorio.ufba.br%2Fri%2Fbitstream%2Fri%2F12005%2F1%2Fo-professor-e-a- educacao-inclusiva.pdf&usg=AFQjCNFXXdewwfUwsl-72BWtDBN_iDWEXw> Accessed April 22, 2017.
SALAZAR F.S. **El conocimiento pedagógico del contenido como categoria de estúdio de La formacion docente.** Revista Electrónica "Actualidades Investigativas en Educación" E- ISSN: 1409-4703 Universidad de Costa Rica Costa Rica Available at < http://www.redalyc.org/articulo.oa?id=44750211> Accessed on: April 15, 2017.
SALES, E. R. **Visualization in mathematics teaching**: an experience with deaf

students. 2014.135f. PhD thesis in Mathematics Education - Universidade Estadual Paulista, Rio Claro, 2013.
SALES, E. R. **Reflecting in silence:** a study of learning to solve additive problems with deaf students and hearing researchers. 2009. 162 f. Master's dissertation in Science and Mathematics Education - Federal University of Pará, Belém, 2008.
SALES, L. M. **Digital technologies in deaf math education in a regular public school:** possibilities and limits. 2010. 113f. Master's dissertation in Education - Pontifical Catholic University of Minas Gerais, Belo Horizonte, 2009.
SANTOS, A.R. and TELLES M.M. **Salamanca Declaration and Inclusive Education**. In: 3° SIMPOSIO EDUCAQÁO E COMUNICADO. Infoinclusion: possibilities for teaching and learning. International edition. Aracaju. Sep.2012. p 77-87.
SANTOS, H. F. **Symmetry and reflection:** investigations in an Inclusive School. 2013.132f. Master's dissertation in Mathematics Education - Bandeirante University of Sao Paulo, Sao Paulo, 2012.
SANTOS, L. I.S.et.al. Face to **face with Nóvoa**: initial and continuing training, social relevance and challenges for the initial and continuing profession Revista de Letras Norte@mentos - Revista de Estudos Linguisticos e Literarios Issue 10 - p.110. Literary Studies 2012/02. Available at <http: //sinop. unemat.br/ projetos/revi sta/index.php/no rteamentos/article/ view/1048/715> Accessed on April 22, 2017.
SASSAKI, R. K. **Terminology on disability in the age of inclusion**. *Revista Nacional de Reabilitagáo.* Sao Paulo, ano 5, n.24, jan./feb. 2002. **Inclusion Paradigm and its educational implications. Available at:<** http://www.ines.gov.br/wp-content/uploads/2014/04/forum5-old1.pdf > Accessed on: January 13, 2017.
SILVA A.S. **Os saberes docentes para a prática pedagógica de alunos com necessidades educativas especiais na escola regular**, 2014,118 f. - Master's dissertation Universidade Estadual de Feira de Santana, Programa de Pós-Graduação em Educação, Feira de Santana, 2014.
SILVA, E.L. **Light, camera, action:** adapting a fractions teleclass for deaf audiences. 2015.107f. Master's dissertation in Mathematics Education - Bandeirante Anhanguera University, Sao Paulo, 2014.
SILVA, G.G. **The teaching of matrices:** a mediated challenge for blind and deaf learners. 2013.144f. Master's dissertation in Mathematics Education - Bandeirante University of Sao Paulo, Sao Paulo, 2012.
SILVA, M.C.A. **Numerical writing by bilingual deaf children.** 2009,226f, Master's dissertation in Education for the Teaching of Science and Mathematics - State University of Maringá, Maringá, 2008.
SOUZA, F.R. **Explorations of equivalent fractions by deaf students:** an investigation of the contributions of MusiCALcolorida. 2011.209f. Master's dissertation in Mathematics Education - Bandeirante University of Sao Paulo, Sao Paulo, 2010.
SOUZA, M.C. **Contextos educacionais inclusivos de alunos surdos:** agoes frente à realidade inclusiva **de** professores **de** matemática da educagao básica. 2014.219f. Master's dissertation in Education for the Teaching of Science and Mathematics - State University of Londrina, Londrina, 2013.
SPENASSATO, Débora; GIARETA, Mariane Kneipp. **Inclusion of deaf students in regular education:** Investigation of the didactic methodological proposals developed by math teachers at EENAV High School. In: Encontro Gaúcho de Edcucagao Matemática, p. 10, 2009. Ijuí. Proceedings. Ijuí: EGEM, 2009. Available at: < http://www.projetos.unijui.edu.br/matematica/cd_egem/fscommand/CC/CC_60.pdfAAcessso: 14deabril de 2017.
STROBEL, Karin. **As imagens do outro sobre a cultura surda**. 2. ed. rev. Florianópolis: Ed. daUFSC, 2009.
TARDIF, M.. **Teaching knowledge and professional training.** Translated by

Francisco Pereira de Lima. Publisher: Vozes. Petrópolis, Rio de Janeiro, 2002.
UNESCO, 1994. **Salamanca of Salamanca.** Available at < http://portal.mec.gov.br/seesp/arquivos/pdf/salamanca.pdf> accessed on December 11, 2016.

APPENDIX

APPENDIX 1: INITIAL QUESTIONNAIRE

Dear participant,

I'd like to get to know you a bit better in order to prepare the tasks for our extension course meetings more appropriately. To do this, I would just ask you to answer the questions sincerely.

Name: ____

Education (check as many boxes as necessary):

() Teaching () Higher education () Pedagogy

() Degree in Mathematics: () completed () in progress

Teaching experience:

() none () less than two years ()two to five years () more than five years

Teaching experience with students with special needs:

() I have already received students with special needs (cite):() not yet

Your opinion:

There's a lot of talk these days about inclusion and inclusive schools. How do you understand these terms?

____

____

____

____

Do you know any schools or teachers that have students with special needs? () yes () no

If so, how does he/she deal with the situation?

____

____

If you had a blind or deaf student in your regular class, how would you feel?

____

____

Do you feel prepared to teach math to students with special needs?

____

____

In your opinion,

()It is more appropriate for students with special needs to study in a specialized school, with teachers prepared to assist them and classmates with the same needs.

() it is more appropriate for students with special needs to study in a mainstream school, interacting with peers of their own age group and learning to live with differences.

() other:

In this extension course, we intend to study ways of teaching mathematics to blind and deaf students, but also to collectively build inclusive teaching activities and proposals. What do you think? ____

____

____

____

<u>What would you like to learn in this course?</u>

____
____
____
____
____
____

Leave your suggestions for the course:

____
____
____
____
____
____
____
____
____

Thank you so much!

APPENDIX 2: FREE AND DISCLAIMED CONSENT FORM (student participant)

I, a student on the Mathematics degree course at UFOP, have been invited by Prof. Ana Cristina Ferreira to take part in an extension course aimed at teaching Mathematics to blind and deaf students from the perspective of school inclusion in regular classes. I know that it has the support of the management of this institution.

I have been informed that the course will take place in six 3-hour meetings on Saturdays and that at the end I will receive a certificate confirming my participation. The meetings will involve workshop moments, in which I will experience teaching experiences focused on the manipulation of materials and activities involving videos and information and communication technologies, as well as study, reading and reflection on Inclusive Mathematics Education. At the end of the course I will have the opportunity to present an activity or teaching proposal, developed by me in partnership with colleagues on the course, aimed at teaching mathematics to blind and deaf students in mainstream classes. The course will be coordinated by Professor Ana Cristina and taught by Professors Viviane Cristiane Costa and Marileny Aparecida Martins, who are carrying out their Master's research under her supervision.

I am aware that, as this is a research project, data will be collected through audio and video recordings of some of the course meetings, as well as through interviews, questionnaires and records produced by the participants. All the material collected will be at my disposal and at the disposal of UFOP throughout the study. The information will be saved on a CD and/or DVD which will be kept by Prof. Ana Cristina Ferreira in her office for five years and then destroyed. At the end of the research, I will be able to access the research in its entirety on the website of the Professional Master's program in Mathematics Education (www.ppgedmat.ufop.br).

I have been informed that my participation will not involve any expense and that I can withdraw from the study at any time without this affecting me. In this case, I will take part in the course as normal and will receive a certificate of participation, but information and images relating to me will not be used. Furthermore, my name will not be mentioned in any documents produced in this research. If, for any reason, I wish to clarify any ethical aspect of the project and/or the activities carried out in it, I know that I can contact the researchers or the CEP through the contacts mentioned at the end of this term.

I have been informed about the proposal and I agree to take part in this research. I also authorize audio and video recording, as well as the other forms of data collection mentioned.
Student's signature

Prof. Ana Cristina Ferreira (responsible) anacf@iceb.ufop.br - 3559-1241
Research Ethics Committee - Federal University of Ouro Preto (CEP/UFOP) University Campus - Morro do Cruzeiro - ICEB II - room 29 - cep@propp.ufop.br -(31) 3559-1368/Fax: (31) 3559-1370

FULL CONSENT FORM (teacher participant)

I, a primary school teacher, have been invited by Prof. Ana Cristina Ferreira to take part in an extension course aimed at teaching mathematics to blind and deaf students from the perspective of school inclusion in regular classes. I know that it has the support of the management of this institution. I have been informed that the course will take place in six 3-hour meetings on Saturdays and that at the end I will receive a certificate confirming my participation. The meetings will involve workshop moments, in which I will experience teaching experiences focused on the manipulation of materials and activities involving videos and information and communication technologies, as well as study, reading and reflection on Inclusive Mathematics Education. At the end of the course I will have the opportunity to present an activity or teaching proposal, developed by me in partnership with colleagues on the course, aimed at teaching mathematics to blind and deaf students in mainstream classes. The course will be coordinated by Professor Ana Cristina and taught by Professors Viviane Cristiane Costa and Marileny Aparecida Martins, who are carrying out their Master's research under her supervision.

I am aware that, as this is a research project, data will be collected through audio and video recordings of some of the course meetings, as well as through interviews, questionnaires and records produced by the participants. All the material collected will be at my disposal and at the disposal of UFOP throughout the study. The information will be saved on a CD and/or DVD which will be kept by Prof. Ana Cristina Ferreira in her office for five years and then destroyed. At the end of the research, I will be able to access the research in its entirety on the website of the Professional Master's in Mathematics Education program (www.ppgedmat.ufop.br).

I have been informed that my participation will not involve any expense and that I can withdraw from the study at any time without this affecting me. In this case, I will take part in the course as normal and will receive a certificate of participation, but the information and images relating to me will not be used. Furthermore, my name will not be mentioned in any documents produced in this research. If, for any reason, I wish to clarify any ethical aspect of the project and/or the activities carried out in it, I know that I can contact the researchers or the CEP through the contacts mentioned at the end of this term.

I have been informed about the proposal and I agree to take part in this research. I also authorize audio and video recording, as well as the other forms of data collection mentioned.
Teacher's signature

Prof. Ana Cristina Ferreira (responsible) anacf@iceb.ufop.br - 3559-1241
Research Ethics Committee - Federal University of Ouro Preto (CEP/UFOP)
University Campus - Morro do Cruzeiro - ICEB II - room 29 cep@propp.ufop.br -(31) 3559-136

APPENDIX 3: COURSE EVALUATIONS

As with all the evaluations proposed throughout the course, the participants did not have to identify their answers. Below are all the questions from the course evaluation, together with a transcript of the participants' answers.

1) Think about the work you and your group presented today. Now, answer: a) How did you feel while preparing the work?

We noticed that three participants didn't answer any of the letters in the first question. We believe that these participants didn't answer because they hadn't introduced themselves or weren't part of a group. Of the others, four participants wrote that they felt cautious, attentive and concerned about preparing a truly inclusive activity, in other words, an activity that all the students could do. Three wrote that they felt good or at ease because they had already carried out a similar activity on another occasion. One wrote that he felt very curious, as he had never seen the methodologies covered in the course. The cost of the activity was also mentioned; one participant wrote that he felt concerned about doing an activity using practical, inexpensive and effective material in terms of learning. Only one participant wrote that his biggest difficulty was the mathematical content. Another participant wrote that he tried to put himself in the deaf person's shoes to plan the activity. And another participant wrote: "*I felt that everything is possible if you are open to the new and when you are willing to dialog, learn to teach*" (Final Evaluation, 5° meeting /12/12/2015).

b) What aspects did you consider most important when planning the activity?

One participant left this letter blank. Four considered the inclusion of everyone. Two considered cost and feasibility. And another four considered the adaptation of mathematical concepts based on the characteristics of a deaf student in a creative way. Group interaction when planning the activity was also mentioned. One participant wrote: "*The involvement of colleagues in the group, thinking first 'what' to work on and 'how', organizing in stages, step by step, to reach the goal*" (Final Evaluation, 50th meeting /12/12/2015).

c) Would you take this activity into the classroom if you had the chance? Why?

All the participants who answered this question wrote yes. There were only two types of justification. Seven participants justified themselves by writing that their activity could be worked on by all students, regardless of whether they were deaf or hearing. And five wrote that their activities would arouse the students' interest and learning since they would be working differently from usual, using concrete material in a playful way.

2) Imagine you've been hired to teach math at a school and you find out a month before classes start that you'll be taking on a deaf student.

a) How would you feel?

The opinions didn't differ much, as five participants answered: worried. Four said: unsure. Two answered: challenged. One answered: lost. And one answered: prepared.

b) How would you prepare to receive him?

Twelve respondents said that through study or research, they would promote the interaction of these students, seek information about the particularities of deaf students and which methodologies to use:

> *"I would come up with various activities to start their interaction both with me and with their classmates."*
> *"I would look for information and new methods to work with this student."*
> *"Looking for really inclusive activities."*
> *"I would study the profile and your needs to be prepared."*
> *"I would research and study and talk to the class to plan what my lessons would be like"* (Final Evaluation, 12/12/15).

One participant wrote that he would pay more attention to deaf students. One wrote that he didn't know. Another replied that he would learn sign language.

3) What did you think about teaching mathematics to the deaf before taking this course? (Use the back of the sheet if you wish or need to)

Nine participants wrote that they thought it was impossible or very difficult, but added that if the teacher dedicates themselves and acquires new knowledge, teaching math to the deaf can become possible:

> *"I thought it was impossible, but with your teachings, a lot can be done with dedication."*
> *"I thought it wouldn't be possible, but when I took part in this course. I realized that nothing is impossible, that there are always ways and new concepts that we can use with these students.*
> *"I thought it was very difficult, but I realized that if I prepare myself and seek new knowledge, it becomes easier to teach."*
> *"I thought it would be more difficult, but with the course I realize that perhaps what is needed is to have the will, to research, to try to see the needs of the students."*
> *"I realized that it's not that difficult, you just need the teacher's willpower."*
> *"I thought the challenge was much bigger, but with the course I realized that what is lacking is information and strategic thinking to reach these students, that is, new*

ways of teaching" (Final Evaluation, 12/12/15).

Five participants wrote that they always thought it was possible because they already worked in this way or because they believed in the professionalism of the teacher or the deaf student's family:

> *"_I always thought it was possible, because in my work there is a need for math activities that make sense in a concrete way for deaf students."*
>
> *"_I've always thought it was possible and important that these people live with this need for math all the time in their daily lives, but we just need to think that they have different ways of understanding."*
>
> *"_I've always believed that we teachers have immense creativity and a great desire to truly teach our students, and for that we must always seek knowledge and above all get to know them so that we can help them."*
>
> *"_I have always believed that it would be and is possible, as long as there is good will, affection and research on the part of the teacher. Nobody is born knowing, we learn throughout our lives, we learn a lot as the need arises, as long as we are open to it."*
>
> *"_I've always thought of this as an important issue that deserves a lot of attention from everyone, education professionals and the student's family"* (Final Evaluation, *12/12/15).*

One participant wrote that he had never thought about it until he came to the course. And that he would definitely think about it from now on.

4) Has your participation in this course changed the way you perceive inclusive education in any way? ( ) Has it changed a lot ( ) Has it changed a little ( ) Has it not changed at all Please explain using the back of the sheet.

Ten participants said it had changed a lot. Of these, only three did not explain themselves. The others explained themselves by writing that they had no knowledge of the subject or reaffirmed the importance of the teacher:

> *"_Because they showed that there are different ways of teaching deaf children, I had the chance to see other people's interest in teaching mathematics. The works presented were very interesting and I will use some of them in my math classes."*
>
> *"_Because now I'll know more about how to deal with these things in education."*
>
> *"_It took me out of my comfort zone and made me see that everyone has the capacity to learn, regardless of their physical, intellectual or mental condition."*
>
> *"_I had a different view, because I know cases of disabilities."*
>
> *"We all have specific needs, so in a classroom diversity is a constant challenge, when I'm in this place I try to contemplate all the students' difficulties."*
>
> *"_Because I didn't know how to teach math to deaf students"* (Final Evaluation, 12/12/15).

Of the other five participants, four said it had changed little. Of these, two justified themselves by writing that they had already taken a course on Inclusive Education during their undergraduate studies. Only one participant said that nothing had changed, as he had always understood Inclusive Education as education for all:

> *"_It hasn't changed my understanding, because that's part of my job, but it's very gratifying to realize that more people (especially young people) are trying to change education. Education doesn't need to be inclusive, it needs to be for everyone without distinction!!!"* (Final Evaluation, 12/12/15).

5) Evaluate this extension course, looking for positive aspects and aspects that could be improved (use the back of the sheet if you wish or need to):

The participants used the rectangle on the left to write positive aspects and the rectangle on the right for aspects that could be improved. Nine used positive aspects and aspects that could be improved as the titles of the rectangles.

One participant used only one of the rectangles and another left this question blank. The answers written by the participants are on A5 (p. 135).

6) What did you think of the activities related to teaching mathematics to the deaf? Why?

All the participants answered this question. They used some adjectives to express their opinions. Three participants did not justify their answers. As well as classifying the activities, they took the opportunity to report on how they felt or what they perceived when they carried out the activity:

> *"Cool and interesting, because it was a new and satisfying experience for everyone."*
> *"Excellent, because the materials used can be used by everyone."*
> *"Interesting, because these activities can be taught to everyone."*
> *"Very interesting, because they were well adapted to the reality of the student and the classroom, with student testimonies, participation of the teachers who are in the classroom and the students."*
> *"The ideas were very good, because the proposals were really inclusive and accessible in the preparation."*
> *"The activities were very well done, really designed to overcome the communication barrier, they are interesting and fun and can be worked on with everyone."*
> *"They were very good, there could have been more."*
> *"I found it less difficult, but no less challenging, because not being able to hear made me concentrate more on the activities and the lack of verbal communication made it harder to communicate."*
> *"I found it interesting because we can apply it to the class in general, but at the same time I found it difficult to communicate because I don't know any Libras."*
> *"Well instructed, because the whole time I was attentive to the tips and steps made by the teacher. In addition to the fun activities that hold our attention. "*
> *"All the activities presented showed us that we can work with all kinds of specific needs."*
> *"In general, all the activities were ones we know, the difference was the slight adaptation they underwent, so they were didactic for both types of students."*
> *"I think it needed more work."*
> *"Very enlightened."*
> *"It seemed to me that the deaf have the advantage in mathematics of being able to see, which makes it much easier. But because they can't hear, they also need the same dedication from the teacher when it comes to describing and giving commands in sign language."*

7) What suggestions would you give for improving this extension course?

Three participants left this question blank. One participant reported that he liked everything and that the course was good. Three participants suggested that the course should last longer. Two wrote that we should propose a second course that deepened the first. One participant wrote that the course should teach more about Libras.

The issue of the different educational levels of the participants was mentioned, with one participant saying that the groups formed during the course activities should not be separated by educational level. Other types of disabilities were also mentioned. One participant wrote that it would be important for the course to address other types of disabilities, such as intellectual and motor.

The deaf student was also mentioned. Two participants wrote that we should get more deaf students to give us their opinions on how they learn math. And apply practical activities with the participation of these students in the course. Only one participant referred to the mathematical content covered. She said that the activities should be specific to a grade/year within the curriculum. This is so as not to assume that concepts have already been worked on.

APPENDIX 4: REFLECTIONS PRODUCED IN THE FIRST TWO MEETINGS

Reading the reflections on the tasks proposed in the first two meetings gave us some interesting data. Here is a summary.

<u>1) How did you feel? Please explain.</u>

The feelings were diverse. Three participants wrote that they felt good because they found the explanations clear and objective. Three others felt awkward, but managed to understand everything and enjoyed the activity. Most of the participants also wrote about the difficulty of communicating and understanding and pointed out that when the procedures were written on the board it made it much easier. They thought it was bad not being able to speak. And almost all of them used the word: uncomfortable. Another participant wrote that he felt deaf, without sound stimuli, which in his opinion filled a void. But he added that he felt able to learn and communicate like a hearing person. We had one participant who wrote that it wasn't a feeling, but that he realized that it is possible for deaf and hearing people to communicate without stress.

Mathematics was also described in one answer. One participant wrote that she felt lost because she found it difficult to remember the theories and practice them. Other participants wrote:

> *"_A horrible feeling, a silent, vague and strange world. I wouldn't like to live like that at all."*
>
> *"_I felt anxious to see what would happen."*
>
> *"_ Reflective on the situation of deaf people in the classroom."* (participant responses, meeting, 28/11/15).

2) Did you find this activity difficult? Why?

The participants' answers were very brief. Two participants answered yes and explained that this activity required prior knowledge of the mathematics content covered and that it was complicated not to speak. The others answered no and explained that the video and the writing on the board with the commands were very helpful. They wrote that the instructions were clear and objective. They also stressed the importance of first paying attention and then carrying out the task. One participant wrote that it was a question of habit. At first he found it difficult, but gradually he got used to it and understood the dynamics of the activity.

3) What mathematical concepts could you work on with her?

One participant wrote them all down. Two participants left this question blank. Most of the participants wrote down concepts related to Plane Geometry and Spatial Geometry, such as: geometric figures, angles, lines, properties of quadrilaterals, diagonals, angles, symmetry, internal angles, segments, areas, axes, sides, vertices, dimensions and the names of plane figures. Three participants wrote different concepts in addition to the above. One wrote construction of space. Another wrote fractions and algebra. The third wrote logic and reasoning.

4) Could this activity be useful for students? Explain your answer.

All the participants answered yes. Only one participant did not give an answer. The majority justified their answers by pointing out that using concrete materials to explore visuals can be very important for understanding theories. Three participants wrote that the change in methodology not only encourages students to learn mathematics but can also develop other skills such as concentration, interpretation and motor coordination.

5) What difficulties did you have carrying out the activity without listening? Why?

One participant left this question blank. Three wrote that they had no difficulty and that the secret of the activity was to pay close attention and stay focused so as not to miss any commands. Another participant wrote that his biggest difficulty was dividing his attention between the video, the teacher's instructions and the folding. The others wrote that communication was the biggest obstacle. They pointed out that at times they didn't understand some of the teacher's explanations and that they didn't know how to express themselves in gestures or in Libras to solve their doubts.

6) Could this activity be carried out in a regular classroom with hearing students? Explain your answer.

All the participants answered yes. They wrote that activities that are different from the usual are important for stimulating students' abilities. They recognized that hearing students can have the same difficulties as deaf students. They concluded that a more visual activity can facilitate a hearing student's learning.

7) Do you think it would make it easier for a deaf student to learn? Why?

All the participants said yes. One participant mentioned that when a dynamic involves everyone, learning is even greater. Most of the participants wrote in their justification that visualization can make it easier since deaf people don't hear. They emphasized the importance of applying theory:

> *"_ Yes, because it's a question of immediate contact with the object and applying the theory in a very engaging way. Learning would certainly be more spontaneous and objective. "* (participant's response, meeting, 28/11/15).

In search of more suggestions for planning the meetings, we read the evaluations from the previous meeting. All the participants praised it highly, especially the activity in which they had to put themselves in the shoes of a deaf person. They enjoyed the proposed activities and emphasized how interesting they found our school routine with a deaf student. They wrote again how difficult they found communication, but reported that they were surprised to learn that a teacher can and should use other means of communication besides speech. One participant wrote that he believed that only deaf people could communicate in Libras and that by taking part in the meeting he learned things he didn't know. They asked for more information on Libras, practical activities and for the proposed activities to be sent to them by email.

In view of these reports, we felt comfortable planning another activity along the same lines as the first, but one that dealt with mathematical content other than geometry. So, when we read the initial questionnaires again, one participant wrote that he would like to learn how to teach fractions to a deaf student.

During the previous meeting, we noticed that the participants seemed uncomfortable while we were talking about the presentation they would be giving at the last meeting. So we told our advisor that it would be better if we gave them some written guidelines. So we decided that, in addition to discussing the text that had to be handed in at the last meeting, we would also give them written guidelines for their final course work. And we would do a practical activity with the content of fractions.

APPENDIX 5: GUIDELINES FOR THE FINAL COURSE WORK

On December 12, the last meeting of the course, the group's work will be presented. This work could be a simple task, a more elaborate activity or even a teaching proposal involving several lessons.
The work will be done individually or in groups.
The work must meet the following conditions: a) have as its objective the teaching of a Mathematics topic in Early Childhood Education, Primary Education or Secondary Education; b) propose a task or activity that can be developed by deaf students.Each participant (or group) will hand in a copy of their work and prepare to present it to the other participants in the course in up to 15 minutes.If they wish, they can use the datashow to present their work.
If the work involves any material, it is important to bring it with you to present to your colleagues.
If you want or need help (suggestions, reading, etc.), please write to: acfufop@gmail.com or to vcosta87@yahoo.com.br.

APPENDIX 6: LITERATURE REVIEW: SUMMARIES OF RESEARCH IN CHRONOLOGICAL ORDER

Oliveira (2005) set out to encourage teachers from two special schools to reflect on the teaching of mathematics to 32 deaf people enrolled in elementary school, with an emphasis on the communication barrier between the pairs (teachers and deaf students). It also described and evaluated the activities and results of a case study which used Origami as a resource in Geometry classes for deaf people with hearing teachers[42] . The author used Vygotsky's reflections on thought and language and Sacks' studies on the deaf as a theoretical basis for her research. The data was collected through observations, proposed written activities and reports of experiences with deaf people inside and outside each Special School. The researcher confirmed that the use of Origami in math classes for deaf students, as well as adding mathematical content, can break down the existing communication barrier.

Gil (2007) investigated the training needs of mathematics teachers of deaf students in the final grades of elementary school, with a view to professional development. This action-research involved a regular school, whose aim was to cater for deaf pupils, and which was run by a denominational association. A collaborative group was formed with the math teacher, the art teacher, the IT teacher and the author of the study. The group reflected on the pedagogical practices experienced with the deaf students of each member through a study supported by Imbernón (2000), Schon (1992), Silva (2000), Mazzotta (2000) and other authors, and jointly planned the methodologies of the activities applied, addressing the content of Plane Geometry with interdisciplinarity. The data collection instruments were: field diaries, notes, camera, camcorder and work done by the students. The results showed that most of the teachers surveyed needed professional training in order to work differently in teaching mathematics to deaf students.

Silva (2008) sought to understand the construction of knowledge, the factors and relationships involved, as well as the hypotheses developed by deaf children about numerical writing. The research was carried out in two stages. The first involved collecting data on the organization of the school and the students. The second involved interviews with 11 bilingual deaf children[43] aged between five and nine at a special education school for the deaf. Based on the Piagetian critical clinical method, the results of the research showed that the deaf children developed hypotheses about numerical writing similar to those identified in the hearing children and that the Brazilian Sign Language (Libras) was a fundamental factor in making this knowledge a reality.

Sales (2008) investigated, in a school specializing in deaf education, the evidence generated in the process of teaching with additive problem solving by six deaf students and one deaf and blind student, who communicate in Libras, enrolled in the second grade of elementary school. This was a collaborative action-research project based on deaf education and the reflective teacher. The data was collected through classroom observations, filming of activities and semi-structured interviews. The analysis of the data allowed the author to see that the environment provided by solving additive problems, using Libras, associated with some didactic resources, established a communication channel favorable to collective interaction and the appropriation of mathematical concepts related to the content worked on.

Nascimento (2009) set out to contribute to quality math education for deaf students, from a bilingual perspective, through a course for math teachers, considering that they only received subsidies for working with these students. Six basic education math teachers who worked in public schools took part in this training through meetings. The research for the preparation of the course used the assumptions of the LDB, or Law° 9.394/96, and bibliographical reviews about school inclusion in the context of deaf education. As a continuing education course, the course was structured in four modules. As well as reading about school inclusion in the context of deaf education, questionnaires and practical activities were proposed in the same context. The author found that teachers were dissatisfied with the direction of inclusion, as they claimed to lack preparation to teach

[42] Teachers without hearing loss.
[43] They use Libras to communicate and Portuguese to write.

students with disabilities. However, he found that the course made it possible to assess the importance of continuing training, showing greater productivity when done in groups.

Sales (2009) analyzed the introduction of Digital Information and Communication Technologies (DICTs) in math classes for 12 deaf students (most of whom were adults and workers) in an evening class of 8ª grade (9° year) of elementary school in a municipal public school. Questionnaires, interviews, observation and focus groups were used. The author confirmed that the deaf students were very interested and motivated in using computers, despite the fact that they had difficulty with the content provided in Portuguese and that the room used as a computer lab in the school was inadequate for the proposed activities.

Souza (2010) tried to help deaf students learn the concept of rational numbers. In the first cycle, eight hearing students aged between 12 and 14 and two deaf students aged 19 and 20 took part. The second cycle took place in a municipal school for students with hearing impairments with 11 students enrolled in grade 7ª (8th year of elementary school), aged between 13 and 20. The research started from a constructionist perspective based on the *Design Research* methodology[44] based on the distinction made by Confrey (the world of counting and the world of equipartition), which involved identifying mathematical models expressed in the students' activities. The author used the learning tool MusiCALcolorida[45] to construct, reconstruct and organize mathematical ideas about the concept of equivalent fractions. Based on the *Design Experiment* methodology[46] , the author developed two cycles: the *design* development cycle and the data collection cycle. The data collection cycle used interviews, video recordings, paper records of the activities carried out, comments from the teachers who accompanied the research and all the ways in which the students used to express mathematical content. The results indicated that the incorporation of MusiCALcolor was fundamental for carrying out the proposed tasks, as it offered deaf and hearing students an autonomous way of checking that the results were correct.

Paixâo (2010) investigated the knowledge that teachers developed in order to include deaf students in math classes in mainstream schools. To do this, she used an ethnographic case study. A collaborative group was set up, made up of the teacher/researcher; a Libras interpreter who worked in the class; and three future teachers, students studying for a degree in Mathematics. Once the group was formed, answers were sought in a 4ª grade (5° year) elementary school class in a public school, made up of 20 hearing students and five deaf students. Through meetings, the exchange of experiences between the group members was encouraged, including reading about inclusion and deaf education. The future teachers then began to interact with the 4th grade (50th year) of elementary school, which had deaf students. Through math activities with concrete materials in the teaching of geometry, they tried to overcome the difficulty of communication. Both the meetings and the activities proposed by the future math teachers were based on the thematic axes of teaching knowledge and deaf culture, discussed and analyzed according to studies by various authors.

The data obtained from meetings, activities and observations was collected using a camera, camcorder and logbooks. Analysis of the data showed that knowledge of theory and practice, together with initial training, were fundamental to building a reflective teacher capable of dealing with the different situations they might encounter in their professional life.

Neves (2011) analyzed math teaching situations with the content of multiplicative problems classified based on Huete and Bravo (2006). The research brought together eight deaf students (who communicate in Libras) in a classroom at a special school, so that three math teachers (one deaf and two hearing, who can communicate in Libras) could explain 14 math problems about the multiplicative concept. After monitoring the classes, the researcher found that the deaf math teacher who communicates in Libras, although he had a command of this language, had limitations when it came to interpreting problems written in Portuguese, which compromised the transmission of mathematical knowledge, damaging the teaching and learning process. In the case of the hearing teachers, despite having common characteristics, they showed different performances because of their

---

[44] Adapting a virtual environment according to the context in which it is used.

[45] A computing environment with a calculator, which uses colors and music to represent numbers and operations.

[46] Exploratory methodology that encourages the construction of strategies for teaching and learning.

experiences. The first math teacher, who is a listener and communicates in Libras, due to his initial mathematical training and the process of acquiring Libras, found it difficult to plan appropriate strategies for organizing information and preserving the mathematical content in the process of translating Portuguese into Libras. The second math teacher, who is a listener and communicates in Libras, because he has a degree in Mathematics, is studying Libras and is married to a deaf woman, managed to establish mathematical communication with interactivity, prioritizing the construction of knowledge to insert mathematical content. The data was collected through filming and analyzed from the perspective of the didactic and pedagogical elements present in the actions of the research subjects. The analysis of the data indicated that, in order to be successful in teaching mathematics to the deaf, the mathematics teacher had to structure the methodologies according to three competencies: mathematical knowledge, Libras and Portuguese language:

> Our research reveals that mastery of the triad of mathematical knowledge, Libras and Portuguese is essential for teaching these students. The lack of mastery of one of these three areas of knowledge results in methodological obstacles, contributing to the realization of learning obstacles (NEVES, 2011, p.111).

Conceigao (2012), in a municipal school, with the help of a Libras interpreter, provided subsidies for understanding the algebraic thinking processes of six deaf students in the 9th grade° aged between 18 and 31. After separating the students into pairs and presenting the *Mathsticks* mathematical microworld (virtual environment),[47] proposed activities on the construction of algebraic expressions through *Mathsticks* sessions. As a theoretical foundation, it used Radford's ideas about algebraic thinking and the different types of generalization: algebraic, arithmetic and naive inductions. Two camcorders were used to collect the data. The results showed that interaction with the *Mathsticks* microworld motivated the students to create algebraic generalizations and work with the notion of indeterminate number. According to the author, the students took the opportunity to systematically express their mathematical ideas in visual-spatial forms using Libras and the tools of the microworld.

Santos (2012), in a municipal school, developed mathematics learning situations for five deaf students and three hearing students in an elementary school class of 6ª (70 years). The exploratory research was based on the *Design Experiments* method. The students were divided into pairs (two pairs of deaf students, one pair of hearing students and a mixed pair of deaf and hearing students) to carry out geometry activities involving the contents of symmetry and reflection. The activities were carried out with the help of the Transtaruga microworld[48] . The research was based on Vygotsky's ideas related to mediation and the way in which learners without access to one or other sensory field sought to overcome the absence by using substitute tools. The study was also based on Papert's constructionist vision, especially his belief that an attunement between mathematical representations and aspects of the human body was important for enabling mathematical learning.

The data was collected using Camtasia[49] , three *camcorders* and the screen and *webcam* images from each pair's *notebook.* The data was analyzed in two stages: preliminary analysis and retrospective analysis. The preliminary analysis sought to reflect on the activities carried out and the interactions between the learners in order to plan possible modifications for the next activity. The retrospective analysis sought to identify and understand the processes of mobilization and appropriation of knowledge about symmetry and reflection that occurred during the interaction with the Transtaruga microworld. The results indicated that all the students became involved with the mathematical properties of the symmetry and reflection content due to the interaction of linguistic resources and physical manifestations related to the movements of the computational objects on the computer screen.

Bezerra (2012) studied the main factors that influence the interaction and communication of deaf people when solving mathematical problems in a virtual environment. She encouraged remote discussions of solutions to mathematical problems with seven deaf students from the 1st year of

---

[47] Microworld created to offer alternative ways of building a number sequence and figures.

[48] Dynamic interactive computer environment with a language and computer use adapted to the mathematical content.

[49] Application that allows procedures to be recorded on the computer screen.

secondary school at a special school.° . It also created a discussion forum in the Moodle Virtual Learning Environment (VLE), following the characteristics of the *Design Research* methodology. After three pilot studies with different groups (deaf and hearing), the VLE was used by the deaf students to discuss four mathematical problems of the problem-process type[50] . The VLE also included videos in Libras to facilitate the interpretation of the problem statements. However, the expected interaction between the participants with the aim of discussing possible solutions occurred very timidly despite the researcher's encouragement. Data was collected using the VLE discussion forum and a semi-structured interview with the help of a Libras interpreter. Data analysis was based on Bairral's model (2002, 2007), which classifies discourse typologies and identifies the nature of those that generate a sequence of argumentative interactions. The results showed little interaction between the deaf students involved in the research. The lack of an appropriate language and the acceptance of the VLE as an educational resource were the main difficulties. However, the researcher felt that it was possible to make changes to the VLE in order to improve the way of working in the discussion forum, and that the interactions between the deaf students in the VLE could be enhanced by questioning and critical interventions made by the math teacher.

Silva (2012) investigated, with the help of a Libras interpreter, the role of material tools as a mediating element between the mathematical concept of matrices and four deaf students in a 2nd year high school class at a state school. The research promoted an exploratory study, using the *Design Experiments* methodology (COBB *et al*., 2003), and used the Matrizmat tool[51] to carry out a sequence of activities related to the concept, equality and addition of matrices. At the end of each activity, a report was made, which allowed the tool and the activities to be redesigned, adapting them to the specific needs of each student. The research was based on the work of Vygotsky (2002) on instruments and signs, Oliveira (1999), Rego (2004) on mediation as an intermediary element in a relationship and Hazin and Meira (2004) on Socio-Historical Psychology (SHP). The manipulation of the Matrizmat tool enabled the deaf students to access a representation of matrices and to understand the notions of equality and addition of matrices.

Borges (2013) sought to understand the development of the teaching and learning of mathematics by deaf students who had the support of a Libras interpreter. °She observed 30 mathematics lessons in two classes in the 9th and 10th year of elementary school in two state schools, case 1 and case 2 respectively. In case 1, there were two deaf students and in case 2, two deaf students. In both cases, 12 mathematical activities were applied to deaf and hearing students. In addition to video and audio recordings, field notes and transcripts of the recordings were used to organize the data collected. The analysis of the data was based on Schick, Williams and Kupermintz (2005), which deals with the proper training of the Libras interpreter to work specifically in the classroom. The results showed a lack of interaction between deaf and hearing people in the school environment, a lack of definition of the role of the Libras interpreter, a lack of activities that explored the visual aspect in teaching mathematics and initial and continuing training that did not take into account the inclusion of deaf students. They also showed that deaf students had difficulties interpreting mathematical statements due to a lack of knowledge on the part of teachers and other professionals.

Assis (2013) investigated the extent to which Libras favors the communication of interpretations that integrate rational numbers, in the form of the fraction |. The author, who knows Libras, conducted a study with ten deaf adults who communicate in Libras. The participants, in pairs, carried out activities involving the visualization of fractions. They could discuss, respond and argue in Libras. The theoretical framework used was Vygotsky (1997), who addressed the importance of interaction and communication, and Nunes and Bryant (1997), who addressed the different meanings of fraction representation. The data was collected by filming and analyzed qualitatively, looking at the forms of communication used. Analysis of the data showed that the meanings attributed to the visual representation of fractions interfered with the way they were signaled.

Sales (2013) used the visuality of eight deaf students, who communicate in Libras, in a 50th

[50] Problems used to encourage the development and practice of problem-solving strategies based on acquired knowledge (LEBLANC; PROUDFIT; PUTT, 2007).

[51] Games and activities using plastic boxes to promote teaching and learning.

grade class at a state school, in an attempt to contribute to the teaching and learning of mathematics. With the help of the math teacher and the class's Libras interpreter, an intervention plan was drawn up which proposed geometry activities using visualization. The research used Inclusive School Education as its theoretical framework. The data was recorded in a field notebook and the activities were filmed. Parents, students and the Libras interpreter were interviewed. The results of the research reinforced the importance of establishing a relationship of mutual collaboration between university and school in order to favor the learning of deaf children. The researcher also highlighted the need for further research on the subject.

Souza (2013) was based on the teaching ideas of 15 mathematics teachers from state primary schools in a region that has deaf students who communicate in Libras in their classes. It sought to understand the reality of these inclusive educational contexts and to contribute to debates and studies on the process of including these students in basic education. It used structured interviews containing questions about deafness, deaf culture, relationships with deaf students and Inclusive Education. The transcripts of the interviews were submitted to the procedures and concepts of Content Analysis proposed by Bardin (1977) until they were structured. The responses of the mathematics teachers from the region surveyed made it possible to understand their knowledge of deafness, inclusion policies and educational actions in the inclusive education of deaf students. The research identified the need for greater investment in the continuing education of teachers so that they can collaborate in the construction of inclusive curricula as required by current educational legislation.

Jesus (2014) analyzed the (de)construction[52] of the geometric thinking of a deaf student in year 8° of elementary school in a municipal school with the use of teaching materials. Pedagogical materials were considered to be all the objects and resources used in the research to enhance and help in the construction of this student's geometric thinking. The research took a cultural-historical approach. The classroom, where the deaf student was in the 8th grade, and the Specialized Educational Assistance (AEE) room were used to develop the sequence of activities related to geometric thinking. All the methodologies were based on the Theory of the Formation of Mental Actions by Stages created by Piotr Galperin and supported by Vygotsky's ideas: social relations between the individual and the outside world, and human/world relations mediated by symbolic systems. In addition to the deaf student, the class math teacher, the ESA math teacher, the Libras interpreter and the deaf instructor took part in the research[53] . Data was collected through interviews, questionnaires, logbook entries, photographs, audio recordings, activities, participant observation and observation grids. The results indicated that the deaf student developed autonomy while solving activities during the project and showed that the teaching materials acted as mediators between the deaf student, the teachers and geometric thinking. Therefore, they pointed out that it is interesting to make the math teacher aware of the potential of a deaf student and to consider working together with the Libras interpreter.

Frizzarini (2014) examined the main semiotic representation registers and possible coordination in the teaching and learning of algebra for deaf students who communicate in Libras. To do this, she planned and applied five activities with seven deaf students in the first year of secondary school at a special school. The activities dealt with inequalities according to Duval's (2009) Theory of Semiotic Representation Registers: linguistic representation, algebraic representation and graphic representation. It also analyzed the prior knowledge of three deaf students from a regular school in Barcelona with an education level corresponding to Brazilian high school. The data was collected through written activities, photographs and filming. The results obtained in this research showed that, while the Brazilian deaf translated algebraic expressions into Libras with the support of graphic registers, the Spanish deaf translated algebraic expressions into sign language with the predominant use of typing, i.e. in the same way as written language. While the inclusion process in

[52] "[...] the term 'de-construction' was used because of the different perspectives on the practices carried out, since, in many cases, before constructing new concepts, we first had to deconstruct inadequate concepts, both from the deaf student, the interpreter and the researcher" (JESUS, 2014, p.20).

[53] Deaf professional who monitors the teaching and learning of deaf students in Specialized Educational Assistance (AEE).

Brazil took place bilingually, in Spain orality proved to be a tradition that had a strong influence on the cognitive process of these students.

Silva (2014) discussed and evaluated the accessibility of the material offered in a type of distance learning in Brazil: the Teleclasses of the Telecurso 2000 Program. The research was divided into three stages. In the first, the initial participants were chosen: three deaf adults. Of these, two deaf people studied in a mainstream school and one in a special school. All of them communicated in Libras, had studied Portuguese in written form and were students in higher education. After choosing the participants, the researcher began to apply the material in its original format, covering the content of fractions. In the second stage, he produced adaptations of the Adapted Lecture and the Adapted Workbook based on Vygotsky (Defectology, 1997), Sacks (Deaf Education, 2010) and Nunes (Teaching Rational Numbers, 2012). In the third stage, the adapted material (Lecture and Workbook) was submitted to the evaluation of the final participants: four deaf people who had similar characteristics to the initial participants. The analysis of the written activities, questionnaires, observations and filming highlighted the need to adapt the Telecurso 2000 program's teleclasses in order to promote accessibility to mathematical content. The adaptations made improved the deaf participants' understanding of the content. However, these adaptations were not enough to develop the educational autonomy of these participants to carry out all the proposed activities more independently.

ANNEX 1: CAMILA'S INTERVIEW

Transcript of the interview:
Camila: *My name is Camila. My sign is... (C sign accompanied by a hand movement). And I'm 18 years old.*
Viviane: *How old were you and what grade did you start in?*
Camila: *5ª grade, 12 years old.*
Viviane: *How were you received by the students and teachers?*
Camila: *Everyone accepted.*
Viviane: Have *you managed to make friends?*
Camila: *I couldn't.*
Viviane: *Why?*
Camila: *Because the other students didn't like being with deaf people.*
Viviane: *And in your classes, did you have an interpreter?*
Camila: *I had an interpreter.*
Viviane: *How did the teachers explain the subject to you?*
Camila: *The interpreter explained.*
Viviane: *Only the interpreter explained?*
Camila: *Just the interpreter.*
Viviane: *And your colleagues didn't offer to help?*
Camila: *Only the interpreter helped.*
Viviane: *How did you do the work? Alone?*
Camila: *I used to do it with my mother and family.*
Viviane: Did *you like math at that time?*
Camila: *I didn't like it. Because I needed to learn math. Multiplying, adding, dividing, subtracting... And in the past I didn't know.*
Viviane: *And do you currently enjoy studying mathematics?*
Camila: I *like it very much. I learn math. I can do tests. Adding, multiplying, dividing, subtracting.*
Viviane: *Does the math teacher teach well?*
Camila: *Yes, I like it very much.*
Viviane: Did *you like the change of teacher?*
Camila: *I didn't like the past.*
Viviane: Is it *better now?*
Camila: *It's better now with teacher Viviane. In the past it was difficult.*
Viviane: *Why do you think it's easier now?*
Camila: *Because it was difficult with the other math teacher. Now I like the teacher.*
It's easy to take a math test. The teacher teaches well.
Viviane: *How do you think people should treat a deaf person?*
Camila: *Say hello, good morning. I like that. Chatting a bit.*
Viviane: Why do *you think it's important to study?*
Camila: *To go to university.*
Viviane: *Of what?*
Camila: *Libras.*
Viviane: *Why?*
Camila: *To learn more. To train. Because I really like Libras.*
Viviane: Do *people talk to you at work?*
Camila: *They talk.*
Viviane: *And do they know Libras?*
Camila: *More or less. I help people who want to learn Libras. They need to learn the alphabet.*
Viviane: *Camila, we're done. Thank you so much for your help!*

ANNEX 2: PARTICIPANTS' OPINIONS OF THE COURSE

Positivos	Negativos
• A visão do aluno deficiente sobre a própria situação, Sobre a escola, as aulas. • As oficinas realizadas que se tornam modelos e podem ser aplicadas. • As experiências adquiridas a partir da sensação de ser cego ou surdo, que nos faz nos colocarmos no lugar do outro	• O tempo que foi curto • A interação entre os participante (misturar os grupos).

Ponto Positivo	Ponto a ser melhorado
• Vocês nos mostraram que todos são capazes, independente de ter ou não necessidade especial. • E de que somos todos capazes	O curso deveria ser de mais tempo.

Mais apoio Mais interesse Mais sugestõis	Mais atenção

POSITIVOS	A MELHORAR
• A PROPOSTA • A PARTICIPAÇÃO DOS ALUNOS DA GRADUAÇÃO DE MATEMÁTICA • O CURSO EM GERAL FOI MUITO BOM	• DURAÇÃO DO CURSO • ASPECTOS DA MATEMATICA NO QUE TANGE OS ~~MATÉRIAS~~ CONTEÚDOS MAIS COMPLEXOS. • MAIS PROPOSTAS PARA A MATEMÁTICA PARA OS SURDOS

- Interação e troca de experiências e conhecimentos. - Oportunidade de falar sobre inclusão.	O tempo foi pouco deveríamos transformar estes encontros em momentos de estudos, ou seja grupo de estudos.

Positivos	negativos
* Troca de ideias * Esclarecimento dos obstaculos * entrevistas com alunos com deficiencia	* Falta de interação com os alunos com deficiencia.

positivo	melhorar
- nos colocar no lugar dessas pessoas com necessidades especiais - ter contato com experiencias distintas - discussões	não segregar os grupos: misturar professores com futuros professores, secretárias, supervisores, etc.

Positivos	Negativos
- vivências surdos e cegos - apostilas	- Pouca carga horária - Poucas vagas

Nos mostrou que mesmo tendo necessidades todos somos capazes, basta querer.	

Aspecto +	Aspecto –
- Os participantes serem de várias áreas - Diversidade de atividades - O trabalho de conclusão do curso	- Pouca integração entre os participantes ("panelinhas") - Senti falta de sugestão bibliografia básica - pessoas que não se encontram e visita de pgra apresentar e pouco tempo para desenvolvimento das atividades de conclusão

(+) Uso de materiais concretos; (+) Abertura para o diálogo; (+) Participação dos "cursistas";	A melhorar: ① Vide nº 07 ② Melhor organização das anotações no quadro para que sigam mais organizados no caderno do aluno

	Pontos Melhorados
* As várias formas de abordar a matemática com esses alunos com deficiências. * Aprendi bastante sobre como passar essa matéria para os alunos.	* O curso poderia durar mais tempo.

Printed by Books on Demand GmbH, Norderstedt / Germany